Biomechanical Insights into Osteoporosis

Edited by Abdelwahed Barkaoui

Published in London, United Kingdom

Biomechanical Insights into Osteoporosis
http://dx.doi.org/10.5772/intechopen.100975
Edited by Abdelwahed Barkaoui

Contributors
Tatiana Petelina, Ksenia Avdeeva, Svetlana Bykova, Natalia Musikhina, Liydmila Gapon, Elena Gorbatenko, Anastasiya Shcherbinina, Imane Ait Oumghar, Abdelwahed Barkaoui, Patrick Chabrand, Wang Qingzhong, Junyan Li, Niuniu Yuan, Huizhen Wang, Hesham Hamoud, Nuno Alexandre, Gonçalo Simões, António Martinho Lopes, Bruna Lopes, Patrícia Sousa, Ana Catarina Sousa, Rui Damásio Damásio Alvites, Ana Colette Mauricio, Tarcísio Guerra Guimarães, Eiman Mohammad Shahrour

First published in London, United Kingdom, 2024 by IntechOpen
IntechOpen is the global imprint of INTECHOPEN LIMITED, registered in England and Wales, registration number: 11086078, 5 Princes Gate Court, London, SW7 2QJ, United Kingdom

British Library Cataloguing-in-Publication Data
A catalogue record for this book is available from the British Library

Additional hard and PDF copies can be obtained from orders@intechopen.com

Biomechanical Insights into Osteoporosis
Edited by Abdelwahed Barkaoui
p. cm.
Print ISBN 978-1-83768-014-6
Online ISBN 978-1-83768-015-3
eBook (PDF) ISBN 978-1-83768-016-0

For EU product safety concerns:
IN TECH d.o.o., Prolaz Marije Krucifikse Kozulić 3, 51000 Rijeka, Croatia,
info@intechopen.com or visit our website at intechopen.com.

Meet the editor

Prof. Abdelwahed Barkaoui is an Associate Professor of Mechanical Engineering/Biomechanics at the International University of Rabat, Morocco. He is currently head of the Mechanics and Advanced Materials Department and coordinator of the Living Mechanics MeV research team at the renewable energy and advanced materials laboratory (LERMA Lab). He is also the head of the mechanical discipline at the energy engineering school. His research mainly targets problems in biomechanics, mechanobiology, and biomedical engineering. Dr. Barkaoui is an editorial board member and reviewer for several international scientific journals. He is the author of three books, *Bone Remodeling Process* (Elsevier), *Finite Element Method and Medical Imaging Techniques in Bone Biomechanics* (Wiley), and *FEM Analysis of the Human Knee Joint: A Review* (Springer), and more than fifty journal publications.

Contents

Preface

We are pleased to present *Biomechanical Insights into Osteoporosis*, a comprehensive work that delves deeply into osteoporosis from a biomechanical perspective. This compilation represents a concerted effort by dedicated researchers and experts to meticulously analyze this complex condition.

Osteoporosis, characterized by a reduction in bone mineral density and increased bone fragility, remains a major global health concern. This book focuses on the biomechanical aspect of osteoporosis, an essential facet for a comprehensive understanding of this condition. Contributors to this volume have invested significant effort to shed light on the intricate mechanisms underlying bone fragility.

Our exploration begins with an in-depth analysis of the biomechanical basis of bone fractures and fracture osteosynthesis in small animals. This chapter establishes the foundations of our understanding of osteoporotic biomechanics, providing critical insights for its treatment and prevention.

In another central chapter, we delve into non-glucocorticoid drug-induced osteoporosis, a topic of paramount importance for osteoporosis management. We scrutinize the underlying mechanisms, risk factors, and therapeutic implications in detail.

Our investigation continues by examining common pathogenetic links between vascular remodeling and bone tissue destruction in postmenopausal women with arterial hypertension. This holistic approach broadens our understanding of bone health by integrating various facets.

The book proceeds to explore promising genetic targets for osteoporosis, offering a glimpse into future treatment directions. We also delve into the ongoing research progress regarding the relationship between bone mineral density and bone metabolism biomarkers.

Finally, our inquiry leads us to a study on the impact of diseases and medical treatments on bone mineral density. This complex analysis highlights how various conditions and therapies influence our skeletal health.

We express our gratitude to our authors for their expertise and commitment in crafting this volume. Their contributions are pivotal in shedding light on the multifaceted nature of osteoporosis from a biomechanical standpoint.

We also extend our heartfelt thanks to our readers for their continued interest in this critical area of medical research. We hope that this book provides you with valuable knowledge and inspires new discoveries to enhance our understanding and management of osteoporosis.

Osteoporosis remains a significant health challenge, but through innovative studies and research in biomechanics, we are making strides toward more effective solutions and treatments.

Welcome to the fascinating world of *Biomechanical Insights into Osteoporosis*.

Abdelwahed Barkaoui
Associate Professor,
International University of Rabat,
Rabat, Morocco

Section 1

Biomechanics

Chapter 1

Study on the Impact of Diseases and Medical Treatments on Bone Mineral Density

Imane Ait Oumghar, Abdelwahed Barkaoui and Patrick Chabrand

Abstract

Several diseases and medical treatments have been found to affect bone quality over decades. Bone mass characteristics summarized in bone mineral density (BMD), geometry, microarchitecture, and mechanical properties are the main parameters permitting to assess the quality of bone. Clinically, the diagnosis of bone diseases and the prediction of bone fracture are largely based on the BMD values. Thus, the investigation of how diseases and treatments alter the BMD value is primordial to anticipate additional treatment for the patient. In this chapter, we summarize the main research studies investigating diseases and treatments' effects on bone quality and more specifically on BMD.

Keywords: bone mineral density, osteoporosis, bone diseases, medical treatments, bone remodeling

1. Introduction

Bone quality depends on the structural and material properties of the bone. By increased mortality and healthcare costs due to bone fractures, bone fragility becomes a major health concern [1]. In order to assess bone fragility, the integration of quantity, quality, and turnover of bone factors is necessary. The BMD, quantified using the clinical imaging technique called X-ray radiography, namely, the dual-energy X-ray absorptiometry (DXA), is one of the most important bone factors reflecting the quality of bone in terms of quantity to assess bone fracture risk in osteoporotic patients [2]. Osteoporosis is a biochemical problem characterized by a decrease in bone mass, a deterioration and an alteration of bone tissue microarchitecture, which increases the risk of fractures. Old bone fracture, oxidative stress, age, and menopause are among the main causes of osteoporosis. Giving its various causes, osteoporosis could affect any gender at any age. For women and men, for example, osteoporosis is associated with aging and more specifically, for women, with hormonal deficiency caused by menopause [3]. Meanwhile for children, BMD loss is generally related to diseases' or treatments' use.

A wide variety of diseases and treatments containing toxic agents for bone cause or contribute to osteoporosis development [4]. In this chapter, we will discuss the diseases and treatments inducing BMD abnormal loss or gain and report different studies' findings concerning diseases and treatment association with osteoporosis. Our work would be subdivided into three main parts: (i) BMD generalities; (ii) osteoporosis in women, men, and children; and (iii) diseases and clinical interventions to limit osteoporosis development.

2. BMD overview

2.1 Bone mineral density

Bone mineral density is the most-used parameter for the prediction of fracture risk in adults. According to the World Health Organization (WHO), bone could be subdivided into four groups based on the BMD values' variation. BMD values are obtained based on DXA images. Based on the calculation of T-score, which represents the number of standard deviation (SD) between studied BMD value and the average value of normal bone of adults of the same sex, the quality of bone can be determined. As far as adults are considered, there are some categories that are more susceptible to get osteoporosis than others; hence, physicians highly recommend them to undergo frequent BMD screenings. By way of illustration, we mention men above 50 years of age who have a historical fracture [5–8], women over 65 years of age, and sometimes the younger women who have an elevated risk of fracture [9]. Based on the fact that osteoporosis is a silent disease and since fragile bones are not painful until the occurrence of fracture, an early screening is always beneficial to detect this disease emergence [10].

For adults, a normal BMD value can be easily determined relative to a specific population. Nevertheless, BMD's use remains difficult for children because of the variation of their bone density and structure during growth. Generally, in this period, bone size and mass grow to reach 90% of the peak bone mass at 18 years of age [11]. Therefore, researchers cannot find a particular normal BMD value to evaluate the bone mineral density levels. In addition to age for children, several variables can also influence the BMD normal value, such as gender, body size, pubertal stage, skeletal maturation, and ethnicity [12]. All these limitations make a diagnosis of osteoporosis more complex [13].

2.2 Dual energy X-ray absorptiometry

DXA imaging technic consists of sending x-rays through the human body, which allows the creation of interior body images based upon the variations of material absorption. Among the many advantages of this technic, we can note its short scanning time, its low radiation dose, and its low cost compared to other imaging technics. DXA provides 2D images of the scanned bone, which can help to extract areal bone mineral density (aBMD) in addition to the hip fracture index (HFRI), and to create a 2D model of the zone of interest, which can be used during finite element modeling. DXA-based finite element models arrive to some extent to determine bone strength and bone behavior vis à vis the mechanical loads. According to [14], DXA-based FE models permit to provide up to 74–77% of experimental femoral strength results. However, despite its advantages, DXA has a poor resolution of images, reducing their quality, which makes the distinction between cortical and trabecular bone impossible. Moreover, DXA 2D images avoid bone structure, which affects the accuracy of obtained a BMD values.

3. Osteoporosis associated BMD alteration

Every human being is susceptible to develop osteoporosis especially by living long. The maximum humans' BMD is reached by 20 years of age; then, it starts decreasing naturally and gradually by approximatively 1%/year. The rate of this decrease variates depending on gender and can be accelerated by the use of steroid, lack of calcium and vitamin D, smoking, high alcohol consumption, and cancer treatments [15].

3.1 Osteoporosis in women

Majority of osteoporosis patients are women because of menopause. During the menopausal transition period, levels of estrogen drop, causing a big disruption in the bone remodeling process. This process is a biological event permitting to renew the old bone matrix by resorbing the old one and replacing it with a new one. Specific bone cells are involved in this process, where their behavior is controlled by numerous biochemical substances, including estrogen. Indeed, estrogen stimulates bone formation and inhibits bone resorption; thus, by its drop, bone resorption rate becomes higher, inducing bone loss. Based on this fact, other factors related to hormonal disturbance also induce osteoporosis. For instance, the problem of oligomenorrhea and taking high doses of glucocorticoids may cause the declination of bone density and increase the fracture risk for women [16, 17]. The assessment of osteoporosis in premenopausal women is not based on BMD value alone because of its high dependency to age. Therefore, there are other signs beyond BMD value, like the exposure of glucocorticoid, parathyroidism, hypogonadism, and eating disorder [18], which make osteoporosis diagnosis easier. Moreover, fractures' occurrence in this period can be also an index of postmenopausal osteoporosis in women [19].

3.2 Osteoporosis in men

In contrary to women, the cause of osteoporosis in men is unknown. However, many studies have associated this disease appearance with age and more specifically with low testosterone production [20]. In the studied population of [21], it has been shown that 22% of male over 50 years old has had a T-score equal to or under −2.5 in femoral neck, indicating that elderly men are more susceptible to develop osteoporotic fractures. However, men are less affected by osteoporosis compared to women. As reported by [22], only 6% of the male population over 50 years have a T-score under −2.5 in femoral neck. For men as well as women, testosterone plays an important role in maintaining a healthy bone. It contributes to stimulating bone-forming cell functioning and estrogen production, which, as aforementioned, contribute to bone formation and inhibit bone resorption. Based on this fact, we can conclude that steroid hormone dysregulation is the most influencing factor on bone health, whether for women or men, especially estrogen that induces serious bone degradation for women during menopause.

3.3 Osteoporosis in children

In addition to adults, the risk of developing osteoporosis or having bone fractures is also important in younger patients.

For girls between 11 and 13 years old and boys between 13 and 15 years old, the occurrence of fractures is totally normal. Indeed, bone grows rapidly before the attainment of the peak bone mineral density; thus, during puberty, bone is under-mineralized, which explains the occurrence of fractures in this period independent of developing any bone disease [23, 24]. However, repeated fractures in early age could be one of the causes inducing osteoporosis and the occurrence of frequent fractures in future. In addition to osteoporosis in children, the decrease of BMD can be related to the appearance of some chronic diseases suchlike cerebral palsy and celiac disease [25] or undergoing treatments such as chemotherapy used for children with lymphoblastic leukemia cancer [26, 27].

As shown before, in women's case, estrogen level is one of the biggest promoters of the bone formation phase in the remodeling process. Estrogen level increases due to its secretion by ovaries during the puberty phase for girls, while it circulates in low concentration for boys and girls before their sexual maturity. Thus, the augmentation of estrogen induces longitudinal bone growth during the puberty period [28]. However, a disturbance of estrogen levels could cause critical bone problems as shown by [29], who found that decreasing concentration of estrogen for girls could induce a sideways curvature of the spine known as scoliosis disease.

4. Disease- and treatment-associated bone deterioration

There are several diseases and treatments that destabilize the bone remodeling process and induce abnormal bone deterioration (**Figure 1**) [30]. Each disease and treatment act differently on hormones, which conducts various bone problems. In this

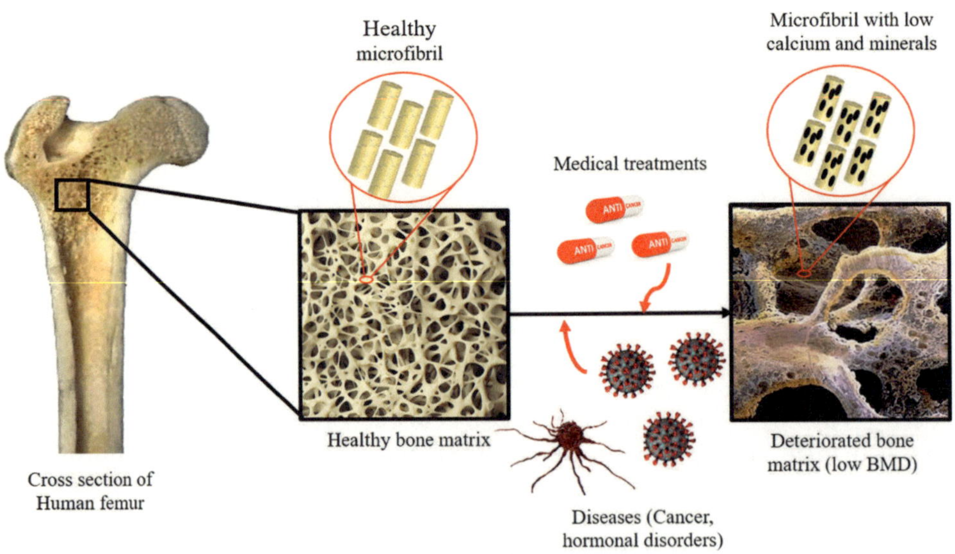

Figure 1.
Illustration of diseases and medical treatments' effect on BMD, bone matrix, and architecture.

section, we are summarizing the most-known diseases and treatments affecting bone and their effect on BMD.

4.1 Disease-associated bone deterioration

Paget disease of bone (PDB): It is a chronic bone disease, considered as the second-most common bone disease after osteoporosis. It is defined as a skeletal growth disorder, leading to bone pain, deformities, deafness, and fractures usually found in the pelvis, spine, femur, and skull [31]. PDB is most common among men than women, and by 50 years of age, the preponderance of being affected by the disease reaches 1 to 5% [32]. The real cause of PDB appearance is not well known. However, many researchers have found a causal relationship between PDB and mutation of the gene encoding sequestosome 1 SQSTM1/p62 that plays an important role in bone-resorbing cell differentiation, activity, and survival [31]. Given this disruption of bone-resorbing cell functioning, the whole bone remodeling process is perturbed, leading to fragile bones and disorganized bone architecture. Otherwise, this disruption causes increased BMD in areas affected by PDB, whether in cortical or trabecular compartments [33].

Osteogenesis imperfecta (OI): Also known as brittle bone disease, OI is a group of rare bone diseases characterized by heterogeneous disturbance of the cognitive tissue. All these diseases are associated with bone mass diminution, increased bone fragility, bone disfigurement, and bone formation insufficiency [34]. OI etiology differs from one disease to another as it depends mainly on the onsets and intensity of each one. Genetic, phenotypic, and functional classifications have been adopted to find out the new causative mutation of OI onset [35]. The most-known OI diseases are X-linked hyposphantaemia, characterized by a mutation of phosphate-regulating endopeptidase that affects vitamin D concentration in the body and consequently bone quality; hypoparathyroidism, characterized by low levels of parathyroid hormone (PTH) that regulates calcium homeostasis; and hypophosphatasia, resulting from the mutation of a responsible gene encoding alkaline phosphatase. The dysregulation of the enzyme disrupts its function to prompt adequate mineralization at an appropriate time in bone tissue. In terms of BMD, OI has been found to be associated with low BMD and also with high BMD in patients with a pathogenic variant of COL1A2, as recently proved by [36].

Myeloma disease (MM): It is a clonal plasma cell proliferative disorder that affects bones by causing bone pain, fractures, and hypercalcemia. It is ranked as the second most-frequent hematological malignancy, with a percentage of 10% [37, 38] in elderly population. This blood cancer is characterized by renal impairment (creatinine >2 mg/dL), hypercalcemia (calcium >11 mg/dL), anemia (hemoglobin <10 mg/dL), the infiltration of clonal plasma with $\geq$60% of the bone marrow, and end organ damage such as lytic lesions in the bone [39, 40]. Because of MM cancer, bone-resorbing cell differentiation is stimulated, and bone-forming cell functioning is suppressed [31]. Due to calcium levels perturbation in blood and bone in addition to bone marrow infiltration, the BMD value is affected.

Breast cancer (BC): It consists of a group of biologically and molecularly heterogeneous diseases that originate from the breast and manifests particularly in the mammary glands. It is the most common type of cancer affecting women worldwide [41]. Whether in its primary state or after metastasizing into the bone, BC affects bones. In a recent research study [42], authors have mimicked bone adaptation during the breast cancer premetastatic state by analyzing the effect of the tumor conditioned media (TCM) of the breast cancer cell line MDA-MB-231 on mice bone. Based on their experiments, they found that TCM injection into mice induced an increase of bone formation characterized by increased bone mineral apposition and altered bone quality such as high mineralization rate, less bone matrix with more carbonate substitution, and disoriented deposition of minerals. Those findings could explain the results of [43] that found a higher BMD in women newly diagnosed with breast cancer. Otherwise, metastatic BC causes osteolytic lesions due to their stimulation of bone-resorbing cell functioning and repression of the forming cell functioning. The tumor cell and bone cell interactions cause a decrease in bone mass, which leads to less bone mineral density in the matrix [44].

Prostate cancer (PC): This disease consists of a cancer that starts in the gland cells of the prostate. The prostate is a small glandular organ where the seminal fluid is produced. PC has a tendency to spread in surrounding organs such as rectum and seminal vesicles and also to distant organs such as bones through lymphatic and hematogenous routes [45]. Its characterization is tightly related to androgenic hormone signaling such as testosterone [46]. As well as estrogen, testosterone plays an important role in bone remodeling regulation. By its exposure in men with PC, it acts on the bone remodeling process directly or indirectly via aromatization to estradiol, leading to enhanced bone formation and repressed resorption [47]. Probably, because of testosterone increased action, PC patients frequently develop osteoblastic lesions [48]. However, PC is still a risk factor for osteoporosis development, leading to decrease BMD value as shown in the retrospective study of Kwon et al. [49].

Lung cancer (LC): This cancer refers to tumors originating in the lung, which is a spongy organ located in the chest. It is caused due to genetic mutations and disruption in protein synthesis resulting generally from individual cigarette smoking. Those DNA mutations disrupt cells' cycle and promote carcinogenesis [50]. In addition to lung complications, patients with this type of carcinoma develop skeletal complications and fractures. In the work of [51] studying the effect of lung cancer cells on bone cells in mice, authors have found that lung tumor cells secrete inhibitory factors that act on bone-reforming cells and inhibit their mineralization, which induce bone loss and decreased BMD.

Cushing disease: It is an endocrine disorder characterized by a hypersecretion of the adrenocorticotropic hormone (ACTH) by the anterior pituitary, leading to an excessive production of cortisol by the adrenal glands [52]. As well as the abovementioned diseases, Cushing disease is linked with bone degradation. Different factors contribute to bone loss and decreased BMD in patients with Cushing syndrome, including a direct effect of the high secreted glucocorticoids on bone cells, enhanced bone resorption, impaired bone formation, and limited calcium absorption [53].

Hypogonadism: It refers to a failure in the gonads' functioning activity. The gonads are testes for men and ovaries for women; both are the principle organs producing sexual hormones such like testosterone, estrogen, and progesterone and gametes, notably eggs and sperm [54]. This disease could arise from Klinefelter syndrome, Kallman syndrome, pituitary disorders, cancer treatment, obesity, aging, and stress [55]. Given that hypogonadism is related to a failure in sexual hormone synthesis, it is also associated with reduction of BMD, especially if sex steroid deficiency occurs at a young age [56]. For males, for example, a low testosterone level is the main cause of low BMD and increased bone fracture risk for patients with primary and secondary hypogonadism [57].

Hyperparathyroidism: It refers to an endocrine disorder characterized by an abnormal elevation of PTH production by parathyroid glands. This causes hypercalcemia by inducing a loss of calcium from the bones and an excessive gain of calcium in the blood. For that reason, hyperparathyroidism induces serious renal and skeletal problems [58]. Due to calcium deficiency in the bone, hyperparathyroidism has been shown to be associated with increased osteoporosis risk and significantly reduced BMD [58].

4.2 Medical treatment-associated bone deterioration

Glucocorticoids: They are a sort of drugs composed of primary stress hormones that regulate various physiological processes [59]. They are used as anti-inflammatory, antiallergic, and immunosuppressive to treat different disease types and more often inflammatory skin disorders such like allergic contact eczema and toxic-irritative eczema [60, 61]. Given its nature, an excess of glucocorticoids leads to reduced calcium intestinal absorption and calcium renal excretion and sex steroid level decrease, which affects bone remodeling by stimulating bone repression and inhibiting formation, leading to bone loss, low BMD, and elevated risk of fracture [62].

Radiotherapy: It consists of using X-rays or subatomic particles directly on tumor cells in curative and palliative settings [63]. Thanks to this procedure, cancer cells shrink, and the local cancer recurrence is reduced [64]. However, RT noticeably affects bone quality, leading to fragility in the zones surrounding the treated tumor area. Indeed, RT decreases cellularity; affects bone-forming cell differentiation, proliferation, and production; and stimulates bone-resorbing cell differentiation using low radiation doses [65]. This explains the significant decrease in BMD value in the area treated with RT as noticed in [66] study investigating RT effect on lumbar vertebrae in women with cervical cancer [66].

Chemotherapy: It involves the use of chemical agents capable of destroying cancer cells by either affecting their macromolecular synthesis by interfering with their DNA or affecting the appropriate functioning of the preformed molecule. This limits cell proliferation and consequently their invasion. Besides their good efficacy on cancer cells, chemotherapy has toxic effects on normal cells, including bone cells, which affects bone quality [67]. Ovarian failure and negative and positive effects on osteoblastogenesis and osteoclastogenesis, respectively, due to chemotherapy induce bone loss [68, 69] in addition to loss of bone mineral density as found in (Bone mineral density change during adjuvant chemotherapy in pediatric osteosarcoma), where children under chemotherapy have shown a decrease of BMD in lumbar spine and femoral neck especially at the end of the therapy [70].

Aromatase Inhibitors (AIs): These are medications used for the treatment of breast cancer, notably estrogen receptor positive, by blocking estrogen synthesis [71]. Because of their action on estrogen, women undergoing AI develop osteoporosis, which induces bone fractures and leads to up to 2.6% loss of BMD per year in lumbar spine [69].

Androgen deprivation therapy (ADT): As well as AIs, ADT is also a treatment controlling sex hormone synthesis, more precisely suppressing or blocking the production or action of male hormones, notably testosterone. Given the important role of testosterone already mentioned, ADT induces osteoporosis. Its effect is much higher than that of AIs as it has been found that ADT causes 4.6% loss of BMD per year in lumbar spine, while AIs induce approximatively 2.6% of BMD decrease [69].

Bone marrow transplant: It is a procedure in which heathy hematopoietic stem cells are administrated into the patient to replace the depleted bone marrow with dysfunctional cells that have been destroyed by treatments such as radiation or high doses of chemotherapy. This permits to supplement bone marrow functioning and to destroy treated malignant tumor cells. The healthy stem cells come from the bone marrow of the patient or from a donor [72]. The hematopoietic cell transplantation has also been identified as one of the medical procedures causing reduced BMD. The major causes of this loss are low sexual hormones notably estrogen and testosterone, secondary hyperparathyroidism due to reduced calcium concentration, and post-transplant steroid therapy—50–60% of patients undergoing this therapy undergo bone loss [73]. In the work of [74], authors have noticed a significant loss in BMD value in the hip, reaching 4.2% for men and women under observation over 1 year of treatment.

5. Medical treatments and dietary recommendation for BMD alteration

The goal of pharmacological therapy is to reduce the risk of fractures by acting on the main agents controlling bone remodeling, which has been affected by osteoporosis. The majority of osteoporosis treatment is antiresorptive (e.g., denosumab, bisphosphonates, estrogen agonist and antagonist, estrogens, and calcitonin), which reduces bone resorption. Otherwise, some anabolic treatments have been developed (e.g., teriparatide), which are dedicated for stimulating bone formation instead of repressing bone-resorbing cell functioning [75]. Both antiresorptive and anabolic agents improve BMD but with variated intensities and can be used for the treatment of one patient. In the work of [76], authors found that the treatment sequence played an important role in osteoporosis treatment. Based on their observation, they have concluded that initiating a treatment by anabolic treatment first followed by a potent antiresorptive treatment is more effective in improving BMD value. In contrary, using antiresorptive agents first, then when the BMD does not sufficiently increase, clinicals suggest to switch to teriparatide, which is not optimal utilization of anabolic treatments.

Besides medications, dietary habits contribute also to improve or deteriorate the bone. Vitamin D and calcium form part of the bone mineral matrix, assuring its strength, and one of the best ways to reach adequate intake of calcium and vitamin D is through healthy eating habits [77]. In a recent meta-analysis of the influence of foods rich in vitamin D on serum 25(OH)D levels, results have shown that those aliments lead to significant increase in serum 25(OH)D as well as BMD [78]. Together with vitamin D, increased calcium intake has been proven to improve BMD by 0.6–1% over 1 year [79]. However, in a recent study [80], dietary calcium intake has not show any association with BMD values in normal participants over 50 years of age, while it has shown a positive change in BMD in women undergoing osteoporosis medical treatment.

6. Conclusion

To maintain its multiple functions, bone is renewed periodically, assuring resorption of old bones and formation of new matrix. However, naturally with age, the process of bone renewing is altered with higher bone resorption and decreased matrix deposition, which lead to decreased BMD and increased fracture occurrence. Several diseases and treatments contribute to or engender osteoporosis or other bone problems. Those diseases and treatments generally affect hormones or bone mineral levels. The most important hormones for a normal bone remodeling process are: (i) estrogen, (ii) testosterone, and (iii) cortisol. And the most important mineral disturbed is calcium. When the absorption of calcium by the bones is inhibited, bones deteriorate easily and undergo fractures. In order to limit the side effects of treatments and additional bad impacts of diseases, using anabolic and antiresorptive treatment is mandatory accompanied by good dietary habits.

Acknowledgements

This work is supported by Partenariat Hubert Curien Franco-Moroccan TOUBKAL (PHC Toubkal) and by the International University of Rabat (UIR).

Conflict of interest

The authors declare no conflict of interest.

Author details

Imane Ait Oumghar[1,2]*, Abdelwahed Barkaoui[1] and Patrick Chabrand[2]

1 LERMA Laboratory, International University of Rabat, El Jadida, Morocco

2 University of Aix-Marseille, ISM, Marseille Cedex, France

*Address all correspondence to: imane.aitoumghar@uir.ac.ma

References

[1] Fonseca H, Moreira-Gonçalves D, Coriolano HJA, Duarte JA. Bone quality: The determinants of bone strength and fragility. Sports Medicine. 2014;**44**(1):37-53. DOI: 10.1007/S40279-013-0100-7

[2] Barkaoui A, Ait Oumghar I, Ben Kahla R. Review on the use of medical imaging in orthopedic biomechanics: Finite element studies. Computer Methods in Biomechanics and Biomedical Engineering Imaging & Visualization. 2021;**9**(5):535-554. DOI: 10.1080/21681163.2021.1888317

[3] Watts NB. Postmenopausal osteoporosis: A clinical review. Journal of Women's Health (2002). 2018;**27**(9):1093-1096. DOI: 10.1089/JWH.2017.6706

[4] Rockville. Bone health and osteoporosis: A report of the surgeon general. 2004

[5] Camacho PM et al. American association of clinical endocrinologists and American college of endocrinology clinical practice guidelines for the diagnosis and treatment of postmenopausal osteoporosis - 2016. Endocrine practice: Official Journal of the American College of Endocrinology and the American Association of Clinical Endocrinologists. 2016;**22**(Suppl 4):1-42. DOI: 10.4158/EP161435.GL

[6] Cosman F et al. Clinician's guide to prevention and treatment of osteoporosis. Osteoporosis international: A Journal established as result of cooperation between the European Foundation for Osteoporosis and the National Osteoporosis Foundation of the USA; 2014;**25**(10):2359-2381. DOI: 10.1007/s00198-014-2794-2

[7] Jeremiah MP, Unwin BK, Greenawald MH, Casiano VE. Diagnosis and management of osteoporosis. American Family Physician. 2015;**92**(4):261-268. DOI: 10.1093/innovait/ins123

[8] Bonnick SL, Harris ST, Kendler DL, McClung MR, Silverman SL. Management of osteoporosis in postmenopausal women: 2010 position statement of The North American Menopause Society. Menopause: The journal of the North American Menopause Society. 2010;**17**(1):25-54. DOI: 10.1097/gme.0b013e3181c617e6

[9] U. S. P. S. T. Force and others. Final recommendation statement: Osteoporosis: Screening. April 2016. 2017

[10] Barr RJ, Stewart A, Torgerson DJ, Reid DM. Population screening for osteoporosis risk: A randomised control trial of medication use and fracture risk. Osteoporosis international: A Journal established as result of cooperation between the European Foundation for Osteoporosis and the National Osteoporosis Foundation of the USA. 2010;**21**(4):561-568. DOI: 10.1007/s00198-009-1007-x

[11] Bailey DA, McKay HA, Mirwald RL, Crocker PRE, Faulkner RA. A six-year longitudinal study of the relationship of physical activity to bone mineral accrual in growing children: The university of Saskatchewan bone mineral accrual study. Journal of the American Society for Bone and Mineral Research. 1999;**14**(10):1672-1679. DOI: 10.1359/jbmr.1999.14.10.1672

[12] Bianchi ML. Osteoporosis in children and adolescents. Bone. 2007;**41**(4):486-495. DOI: 10.1016/j.bone.2007.07.008

[13] Jones G, Ma D, Cameron F. Bone density interpretation and relevance in

Caucasian children aged 9-17 years of age: Insights from a population-based fracture study. Journal of the International Society for Clinical Densitometry. 2006;**9**(2):202-209. DOI: 10.1016/j.jocd.2006.02.004

[14] Dall'Ara E, Eastell R, Viceconti M, Pahr D, Yang L. Experimental validation of DXA-based finite element models for prediction of femoral strength. Journal of the Mechanical Behavior of Biomedical Materials. 2016;**63**:17-25. DOI: 10.1016/j.jmbbm.2016.06.004

[15] Wilson DJ. Osteoporosis and sport. European Journal of Radiology. Jan 2019;**110**:169-174. DOI: 10.1016/J.EJRAD.2018.11.010

[16] Cohen S et al. Risedronate therapy prevents corticosteroid-induced bone loss: a twelve-month, multicenter, randomized, double-blind, placebo-controlled, parallel-group study. Arthritis and Rheumatism. 1999;**42**(11):2309-2318. DOI: 10.1002/1529-0131(199911)42:11<2309::AID-ANR8>3.0.CO;2-K

[17] Saag KG et al. Alendronate for the prevention and treatment of glucocorticoid-induced osteoporosis. Glucocorticoid-Induced Osteoporosis Intervention Study Group. The New England Journal of Medicine. 1998;**339**(5):292-299. DOI: 10.1056/nejm199807303390502

[18] Leib E. Diagnosis of osteoporosis in men, premenopausal women, and children. Journal of the International Society for Clinical Densitometry. 2004;**7**(1):17-26. DOI: 10.1385/JCD:7:1:17

[19] Wu F et al. Fractures between the ages of 20 and 50 years increase women's risk of subsequent fractures. Archives of Internal Medicine. 2002;**162**(1):33-36. DOI: 10.1001/archinte.162.1.33

[20] Mohamad NV, Soelaiman IN, Chin KY. A concise review of testosterone and bone health. Clinical Interventions in Aging. Sep 2016;**11**:1317. DOI: 10.2147/CIA.S115472

[21] Melton LJ, Atkinson EJ, O'Connor MK, O'Fallon WM, Riggs BL. Bone density and fracture risk in men. Journal of Bone and Mineral Research. 1998;**13**(12):1915-1923. DOI: 10.1359/jbmr.1998.13.12.1915

[22] Looker AC et al. Prevalence of low femoral bone density in older U.S. adults from NHANES III. Journal of Bone and Mineral Research. 1997;12(11):1761-1768. DOI: 10.1359/jbmr.1997.12.11.1761

[23] Landin LA. Fracture patterns in children. Analysis of 8,682 fractures with special reference to incidence, etiology and secular changes in a Swedish urban population 1950-1979. Acta Orthopaedica Scandinavica. Supplementum. 1983;**202**:1-109. DOI: 10.3109/17453678309155630

[24] Bailey DA, Wedge JH, McCulloch RG, Martin AD, Bernhardson SC. Epidemiology of fractures of the distal end of the radius in children as associated with growth. The Journal of Bone and Joint Surgery. American. 1989;**71**(8):1225-1231. DOI:10.2106/00004623-198971080-00016

[25] Schmidt OS. Pathologic fractures in children with cerebral palsy. Orthopedic Management of Children with Cerebral Palsy: A Comprehensive Approach. 2015;**26**(5):303-310

[26] Van Der Sluis IM, Van Den Heuvel-Eibrink MM, Hählen K, Krenning EP, De Muinck Keizer-Schrama SMPF. Bone mineral density, body composition, and height in long-term survivors of acute lymphoblastic leukemia in childhood. Medical and Pediatric Oncology. 2000;**35**(4):415-420. DOI: 10.1002/1096-911X(20001001)35:4<415::AID-MPO4>3.0.CO;2-9

[27] Van Der Sluis IM, Van Den Heuvel-Eibrink MM, Hählen K, Krenning EP, De Muinck Keizer-Schrama SMPF. Altered bone mineral density and body composition, and increased fracture risk in childhood acute lymphoblastic leukemia. The Journal of Pediatrics. 2000. DOI: 10.1067/mpd.2002.125728

[28] Szyska-Skrobot D et al. Free and bioavailable fractions of sex steroids may influence bones in young men, depending on age and oestradiol level. Endokrynologia Polska. 2014;**65**(5):357-364. DOI: 10.5603/EP.2014.0049

[29] Kulis A et al. Participation of sex hormones in multifactorial pathogenesis of adolescent idiopathic scoliosis. International Orthopaedics. 2015;**39**(6):1227-1236. DOI: 10.1007/s00264-015-2742-6

[30] Ait Oumghar I, Barkaoui A, Chabrand P. Mechanobiological behavior of a pathological bone. In: Biomechanics and Functional Tissue Engineering. London, UK: Intechopen; Nov 2021. DOI: 10.5772/intechopen.97029

[31] Ait Oumghar I, Barkaoui A, Chabrand P. Toward a mathematical modeling of diseases' impact on bone remodeling. Technical Review, Frontiers in Bioengineering and Biotechnology. 2 Nov 2020;**8**(584198):1236. DOI: 10.3389/fbioe.2020.584198

[32] Appelman-Dijkstra NM, Papapoulos SE. Paget's disease of bone. Best Practice & Research. Clinical Endocrinology & Metabolism. 2018;**32**(5):657-668. DOI: 10.1016/j.beem.2018.05.005

[33] Cherian RA, Haddaway MJ, Davie MWJ, McCall IW, Cassar-Pullicino VN. Effect of Paget's disease of bone on areal lumbar spine bone mineral density measured by DXA, and density of cortical and trabecular bone measured by quantitative CT. The British Journal of Radiology. 2000;**73**(871):720-726. DOI: 10.1259/BJR.73.871.11089462

[34] Forlino A, Marini JC. Osteogenesis imperfecta. The Lancet. (London, England) 2016;**387**(10028):1657-1671. DOI: 10.1016/S0140-6736(15)00728-X

[35] Bacon S, Crowley R. Developments in rare bone diseases and mineral disorders. Therapeutic Advances in Chronic Disease. 2018;9(1):51-60. DOI: 10.1177/2040622317739538

[36] Graves LE et al. High bone mineral density osteogenesis imperfecta in a family with a novel pathogenic variant in COL1A2. Hormone Research in Pædiatrics. 2020;**93**(4):263-271. DOI: 10.1159/000510463

[37] Greenlee RT, Murray T, Bolden S, Wingo PA. Cancer statistics, 2000, CA: A cancer. Journal for Clinicians. 2000;**50**(1):7-33. DOI: 10.3322/canjclin.50.1.7

[38] Palumbo A, Rajkumar SV. Treatment of newly diagnosed myeloma. Leukemia. 2009;**23**(3):449-456. DOI: 10.1038/leu.2008.325

[39] Rajkumar SV, Kumar S. Multiple myeloma: Diagnosis and treatment. Mayo Clinic Proceedings. 2016;**91**(1):101-109. DOI: 10.1016/j.mayocp.2015.11.007

[40] Kyle RA, Rajkumar SV. Criteria for diagnosis, staging, risk stratification and response assessment of multiple myeloma. Leukemia. 2009;**23**(1):3-9. DOI: 10.1038/leu.2008.291

[41] Feng Y et al. Breast cancer development and progression: Risk factors, cancer stem cells, signaling pathways, genomics, and molecular pathogenesis.

Genes and Diseases. 2018;**5**(2):77-106. DOI: 10.1016/J.GENDIS.2018.05.001

[42] Chiou AE, Liu C, Moreno-Jiménez I, Tang T, Wagermaier W, Dean MN. et al. Breast cancer-secreted factors perturb murine bone growth in regions prone to metastasis. Science advances. 2021;7(12):eabf2283. DOI: 10.1126/sciadv.abf2283

[43] Fraenkel M et al. Bone mineral density in women newly diagnosed with breast cancer: A prospective cohort study. NPJ Breast Cancer. 17 Feb 2022;**8**(1):21. DOI: 10.1038/S41523-022-00388-Z

[44] Gregory LS, Choi W, Burke L, Clements JA. Breast cancer cells induce osteolytic bone lesions in vivo through a reduction in osteoblast activity in mice. PLoS One. 2013;**8**(9):e68103. DOI: 10.1371/journal.pone.0068103

[45] Rosario E, Rosario DJ. Localized Prostate Cancer - StatPearls - NCBI Bookshelf, 2022. Available from: https://www.ncbi.nlm.nih.gov/books/NBK563248/. [accessed Aug. 22, 2022]

[46] Murray TBJ. The pathogenesis of prostate cancer. Prostate Cancer. May 2021:29-42. DOI: 10.36255/exonpublications.prostatecancer.pathogenesis.2021

[47] Clarke BL, Khosla S. Androgens and bone. Steroids. 2009;**74**(3):296. DOI: 10.1016/j.steroids.2008.10.003

[48] Lin SC, Yu-Lee LY, Lin SH. Osteoblastic factors in prostate cancer bone metastasis. Current Osteoporosis Reports. 2018;**16**(6):642. DOI: 10.1007/S11914-018-0480-6

[49] Kwon T et al. Bone mineral density in prostate cancer: A comparative study of patients with prostate cancer and healthy controls using propensity score matching. Urology. 2014;**83**(2):385-392. DOI: 10.1016/j.urology.2013.08.045

[50] Siddiqui F, Vaqar S, Siddiqui AH. Lung Cancer. In: Cambridge Handbook of Psychology, Health and Medicine. 2nd ed. StatPearls, StatPearls Publishing. 5 Dec 2022. pp. 605-606. DOI: 10.1017/cbo9780511543579.138

[51] Berent TE, Dorschner JM, Craig TA, Drake MT, Westendorf JJ, Kumar R, Lung tumor cells inhibit bone mineralization and osteoblast activity. Biochemical and Biophysical Research Communications. Nov 2019;**519**(3):566. DOI: 10.1016/J.BBRC.2019.09.045

[52] Kairys N, Schwell A, Haddad LM. Cushing disease (nursing), StatPearls. 2021. [Accessed: Aug. 19, 2022]. [Online]. Available from: http://www.ncbi.nlm.nih.gov/pubmed/33760467

[53] Lodish MB et al. Effects of cushing disease on bone mineral density in a pediatric population. The Journal of Pediatrics. 2010;**156**(6):1001. DOI: 10.1016/j.jpeds.2009.12.027

[54] Kumar P, Kumar N, Thakur DS, Patidar A. Male hypogonadism: Symptoms and treatment. Journal of Advanced Pharmaceutical Technology & Research. 2010;**1**(3):297. DOI: 10.4103/0110-5558.72420

[55] Sizar O, Schwartz J. Hypogonadism. In: Urology at a Glance. StatPearls, StatPearls Publishing; 27 Jun 2022. DOI: 10.1007/978-3-642-54859-8_21

[56] Da Costa JV, Pereira-Lima JFS, Da Costa Oliveira M. Bone mineral density in early-onset hypogonadism and the effect of hormonal replacement. Journal of Clinical Densitometry. 2004;7(3):334-340. DOI: 10.1385/JCD:7:3:334

[57] Golds G, Houdek D, Arnason T. Male hypogonadism and osteoporosis: the effects, clinical consequences, and

treatment of testosterone deficiency in bone health. International Journal of Endocrinology. 2017;**2017**. DOI: 10.1155/2017/4602129

[58] Pokhrel B, Leslie SW, Levine SN. Primary hyperparathyroidism, StatPearls. Jun 2022 [Accessed: Aug. 19, 2022]. [Online]. Available from: https://www.ncbi.nlm.nih.gov/books/NBK441895/

[59] Hodgens A, Sharman T. Corticosteroids - StatPearls - NCBI Bookshelf. StatPearls, StatPearls Publishing; 26 Jul 2022. Available from: https://www.ncbi.nlm.nih.gov/books/NBK554612/. [Accessed Aug. 22, 2022]

[60] Chourpiliadis C, Aeddula NR. Physiology, glucocorticoids, StatPearls. Jul 2022. [Online]. Available from: https://www.ncbi.nlm.nih.gov/books/NBK560897/. [Accessed: Aug. 19, 2022]

[61] Niculet E, Bobeica C, Tatu AL. Glucocorticoid-induced skin atrophy: The old and the new. Clinical, Cosmetic and Investigational Dermatology. 2020;**13**:1041. DOI: 10.2147/CCID.S224211

[62] Manelli F, Giustina A. Glucocorticoid-induced osteoporosis. Trends in Endocrinology and Metabolism. 2000;**11**(3):79-85. DOI: 10.1016/S1043-2760(00)00234-4

[63] E. V. Maani and C. V. Maani, Radiation therapy - StatPearls - NCBI bookshelf. StatPearls, StatPearls Publishing; 24 Oct 2021. Available from: https://www.ncbi.nlm.nih.gov/books/NBK537036/. [Accessed Aug. 19, 2022]

[64] Jonathan Yang T, Ho AY. Radiation therapy in the management of breast cancer. Surgical Clinics of North America. 2013;**93**(2):455-471. DOI: 10.1016/j.suc.2013.01.002

[65] Donaubauer AJ, Deloch L, Becker I, Fietkau R, Frey B, Gaipl US. The influence of radiation on bone and bone cells—Differential effects on osteoclasts and osteoblasts. International Journal of Molecular Sciences. 2020;**21**(17):1-19. DOI: 10.3390/IJMS21176377

[66] Wu J et al. Bone mineral density changes within and outside of RT fields used to treat cervical cancer. International Journal of Radiation Oncology Biology Physics. 2021;**111**(3):e625-e626. DOI: 10.1016/j.ijrobp.2021.07.1662

[67] Amjad MT, Chidharla A, Kasi A. Cancer Chemotherapy - StatPearls - NCBI Bookshelf. 2022. Available from: https://www.ncbi.nlm.nih.gov/books/NBK564367/. [accessed Aug. 22, 2022]

[68] Wissing MD. Chemotherapy- and irradiation-induced bone loss in adults with solid tumors. Current Osteoporosis Reports. 2015;**13**(3):140. DOI: 10.1007/S11914-015-0266-Z

[69] Drake MT. Osteoporosis and cancer. Current Osteoporosis Reports. 2013;**11**(3):163. DOI: 10.1007/S11914-013-0154-3

[70] Ahn JH, Cho WH, Lee JA, Kim DH, Seo J-H, Lim JS. Bone mineral density change during adjuvant chemotherapy in pediatric osteosarcoma. Annals of Pediatric Endocrinology & Metabolism. 2015;**20**(3):150-154. DOI: 10.6065/apem.2015.20.3.150

[71] Dhankhar R, Vyas SP, Jain AK, Arora S, Rath G, Goyal AK. Advances in novel drug delivery strategies for breast cancer therapy. Artificial Cells, Blood Substitutes, and Biotechnology. 2010;**38**(5):230-249. DOI: 10.3109/10731199.2010.494578

[72] Khaddour K, Hana CK, Mewawalla P. Hematopoietic Stem Cell Transplantation - StatPearls - NCBI Bookshelf. StatPearls, StatPearls Publishing; 27 Jun 2022.

Available from: https://www.ncbi.nlm.nih.gov/books/NBK536951/. [Accessed Aug. 22, 2022]

[73] Weilbaecher KN. Mechanisms of osteoporosis after hematopoietic cell transplantation. Biology of Blood and Marrow Transplantation. 2000;**6**(2A):165-174. DOI: 10.1016/S1083-8791(00)70039-5

[74] Stern JM et al. Bone density loss after allogeneic hematopoietic stem cell transplantation: A prospective study. Biology of Blood and Marrow Transplantation. 2001;**7**(5):257-264. DOI: 10.1053/bbmt.2001.v7.pm11400947

[75] Tu KN et al. Osteoporosis: A review of treatment options. P T. 2018;**43**(2):92-104

[76] Cosman F, Nieves JW, Dempster DW. Treatment sequence matters: Anabolic and antiresorptive therapy for osteoporosis. Journal of Bone and Mineral Research. 2017;**32**(2):198-202. DOI: 10.1002/jbmr.3051

[77] Muñoz-garach A, García-fontana B, Muñoz-torres M. Nutrients and dietary patterns related to osteoporosis. Nutrients. 2020;**12**(7):1-15. DOI: 10.3390/nu12071986

[78] Tangestani H, Djafarian K, Emamat H, Arabzadegan N, Shab-Bidar S. Efficacy of vitamin D fortified foods on bone mineral density and serum bone biomarkers: A systematic review and meta-analysis of interventional studies. Critical Reviews in Food Science and Nutrition. 2020;**60**(7):1094-1103. DOI: 10.1080/10408398.2018.1558172

[79] Tai V, Leung W, Grey A, Reid IR, Bolland MJ. Calcium intake and bone mineral density: Systematic review and meta-analysis. BMJ. 2015;**351**. DOI: 10.1136/BMJ.H4183

[80] Bristow SM, Bolland MJ, Gamble GD, Leung W, Reid IR. Dietary calcium intake and change in bone mineral density in older adults: A systematic review of longitudinal cohort studies. European Journal of Clinical Nutrition. 2022;**76**(2):196-205. DOI: 10.1038/S41430-021-00957-8

Chapter 2

Biomechanical Basis of Bone Fracture and Fracture Osteosynthesis in Small Animals

Nuno Alexandre, Gonçalo Simões, António Martinho Lopes, Tarcísio Guerra Guimarães, Bruna Lopes, Patrícia Sousa, Ana Catarina Sousa, Rui Damásio Alvites and Ana Colette Maurício

Abstract

The mastery of concepts related to biomechanics in bone fracture is crucial for the surgical success of osteosynthesis. The understanding of the basics of bone fracture is a skill fundamental to the choice of the correct method of osteosynthesis. Deep knowledge of implants, namely, their mechanic characteristics, and the correct technical use following the recommended guidelines for each type are crucial factors to decrease surgical failure and complications rate. This chapter reviews the biomechanical parameters of fracture repair that influence construct stiffness and strength. The authors also provided practical examples of the biomechanics concepts applied in clinical practice during this chapter. Metal alloys used in orthopedic implants are also fundamentally reviewed in their physical properties during this chapter. Fracture patterns vary hugely among patients and contributed to the difficult understanding of forces acting in fracture lines. However, fracture biomechanics basic knowledge and how osteosynthesis methods counteract the forces acting on fractures are key to surgical success.

Keywords: biomechanics, bone tissue, forces, fracture, dog, cat, osteosynthesis

1. Introduction

Biomechanics is a sub-branch of mechanics that studies the concepts of mechanics applied to the musculoskeletal system and the biomaterials used for treating orthopedic diseases. The structure, function, and motion of musculoskeletal tissues and their changes in orthopedic diseases are the main research topics of this science. The basic knowledge of the physical, chemical, and mechanical properties of biomaterials used for producing implants and prostheses is key to the orthopedic surgeon's understanding of why certain materials are used instead of others. The functional (mechanical) performance of implants and prostheses is strictly related to their material composition and design, being therefore the basic knowledge that the orthopedic surgeon should master and potentially influence his surgical planning or clinical decision.

Biomechanics encompasses the traditional branches of mechanics: kinematics, statics, and dynamics.

Kinematics is the study of motion without considering the forces that cause it and includes concepts such as trajectory, velocity, and acceleration. Motion can be a combination of translations and rotations, with translations involving the same displacement vector for all points in the body, while rotations involve different displacement vectors for different points.

Statics characterizes the forces acting on an object at rest or moving at constant velocity with zero acceleration. These forces can be direct forces or moments, which are equal and opposite forces acting on a body separated by a distance. The application of forces and moments to a body changes its state of rest. Equilibrium is a key principle in statics, and a body is in equilibrium when the sum of all applied loads is zero. In joints, applied forces include external loads such as body weight and internal loads such as muscle forces generated to maintain the joint in equilibrium. The equilibrium principle is used to analyze joint loading in a static context, where the joint of interest is studied in isolation from the rest of the body, and all forces and moments acting on it are identified. The resulting joint reaction force is then determined using the equilibrium condition.

Dynamics, a branch of mechanics, is concerned with the effects of forces on an object and the changes they produce in the object's motion. It encompasses the principles of both statics and kinematics by examining the actions of forces and the resulting motion and acceleration of the object. In orthopedic biomechanics, dynamic analysis is frequently utilized for activities such as gait studies. This involves determining the acceleration of body parts at any given time and the forces necessary to create these accelerations. The resulting forces are then determined using static analysis methods to obtain the resulting forces over the desired range of motion.

The interaction of biomaterials with tissues and cells is the ability of a biomaterial to perform its function without eliciting toxic or injurious effects on biological systems and is called biocompatibility, and it influences the mechanical performance of implants/prostheses in the short and long term. Nowadays, the biocompatibility concept includes bioinertia, biofunctionality, and biostability (acute and chronic toxicity of materials to tissues). Biointegration or colonization of implants by neighboring tissues is also framed in the concept of biocompatibility and is an important factor in the long-term biomechanical performance of implants/prostheses that should not be overlooked in clinical decisions.

Synthetic materials mainly metals and their alloys used for implants/prostheses are classified according to their biocompatibility as well as by their mechanical properties, such as tensile, compressive, and shear strength; hardness; stiffness; fatigue resistance to cyclic or acute loading; and creep behavior. The creep concept is a type of metal deformation that occurs at stresses below the yield strength (at elevated temperatures); it defines the stress at which metal begins to plastically deform. Factors such as ease of manufacture, cost, and production quality dictate the potential for the application of a biomaterial in orthopedics.

Load-deformation and stiffness, stress-strain, and elasticity are interconnected concepts to the understanding of the mechanical performance of implants and bone tissue that will be addressed in this chapter.

A thorough understanding of the unique biomechanical properties, characteristics, and behaviors of bone tissue, their alterations in disease, and the implants used in companion animal orthopedic surgery is essential for achieving successful results when attempting to manipulate bone healing. There is a consensus in the field of orthopedic surgery for companion animals that mastering these principles is associated with a low rate of postoperative complications.

A basic understanding of biomechanical principles and biomaterials knowledge is a fundamental skill for the companion animal orthopedic surgeon and forms an important component in the education of surgical trainees.

In this chapter, the information provided is divided into five main areas: biomechanical basic concepts, fracture biomechanics, biomechanics of bone tissue, applied fracture biomechanics to common clinical presentations in small animal osteosynthesis, and biomechanics of implant biomaterials, covering what the authors considered in-depth knowledge of biomechanical principles of bone fracture and applied biomechanics to fracture osteosynthesis in small animals.

The main objective of this chapter is to provide information about biomechanics applied to fracture management in small animals that will help the veterinary surgeon to take more evidence-based decisions with the ultimate goal of surgical success.

2. Biomechanical basic concepts

2.1 Strain

Strain is a local deformation parameter expressed as units of length per length, usually expressed as a percentage, and is therefore dimensionless when the bone is loaded with different force vectors.

The mathematical definition (Eq. (1)) of strain is the change in length divided by the original length.

Formula for strain calculus:

$$\text{Strain (\%)} = \frac{\text{Change in length (mm)}}{\text{Original length (mm)}} \tag{1}$$

Due to the dimensionless characteristic of this parameter, strain provides a clinically useful scaled measure of the displacement of bone fragments and can compare strain values in bones of different lengths. For example, a 1 mm fracture gap displacement is more significant in a 10 mm rat (strain 10%) femur than in a 300-mm dog femur (strain 0.3%) [1, 2].

2.2 Stress

One of the main functions of the appendicular skeleton is to support the body weight in rest or during movement, and consequently, bone is the tissue that supports more mechanical loads. When a force is applied to a bone, this will cause a stress situation [1, 2].

By definition, like pressure, stress is a local force expressed in units of force per unit area (Eq. (2)). The SI unit of force is the Newton, and force is often expressed as N/m^2 or Pascal (Pa) [1, 2].

Formula for stress calculus:

$$\text{Stress}\left(\frac{\text{N}}{\text{mm}^2}\text{orPa}\right) = \frac{\text{Load (N)}}{\text{Cross} - \text{sectional area (mm}^2)} \tag{2}$$

The damage that the load will cause depends on the area over which it is being distributed: a large force applied over a small area will result in greater stress, the contrary not being true. For instance, that is what happens when a skeletally immature dog falls from a height and supports all the weight on the hind limb. In these cases, the load will be equally distributed proximally and distally to the knee; however, due to the minor dimension of the distal part (tibial crest) of the tibial tuberosity, the fracture is normally located in this point due to stress concentration in a small area [3].

Bone tissue is constantly submitted to mechanical loads, comprehending forces/loads of compression (axial), torsion, tension, bending, and shearing. Usually, these forces act in combination; however, they can be more predominant in an isolated way in certain locations of bones. Conceptually, it is considered that the deformation occurs in the bone tissue when small animals move and will vary between 0.04 and 0.3%, hardly exceeding 0.1%. This interval characterizes the elastic deformation of the bone, which is conceptually an initial response to the establishment of a load in the bone. In this scenario, the deformation/length of the bone returns to the initial dimensions/shape once the load is removed. It is a natural and important process for the homeostasis of bone tissue. An interesting characteristic of bone tissue is related to bone deformation according to the mechanical load that is applied. There are materials conceptually defined as isotropic, which respond to mechanical load regardless of the orientation of the material. By contrast, a material is considered anisotropic if the response to mechanical load varies with orientation. Bone is an example of an anisotropic material with mechanical properties that depend on the orientation of the bone lamellae. Thus, the mechanical properties are not equal in all directions and depend on the direction of the load applied. Long bones are stronger in longitudinal orientation than in tangential or radial orientation, since osteons have a longitudinal orientation in cortical bone. The maximum load that the bone will support is directly related to the direction in which the force is being applied. An illustrative example of this statement is that the appendicular bone supports a greater axial (compressive) load if compared to the transverse load. This difference between maximum strength in different directions emphasizes the anisotropic characteristic of bone. An example of an isotropic material is the stainless-Steel 316L (metallic alloy commonly used in the production of plates, which presents a similar behavior regardless of the direction of the load that is applied, with a similar resistance [3].

From a mechanical point of view, bone is also considered a viscoelastic tissue. Viscoelasticity is the property of materials that exhibit both viscous and elastic characteristics when undergoing deformation. Practically, this represents the ability of the bone to resist deformation without loss of definitive structural integrity. Viscoelasticity of bone is dependent on several factors including water, mineral, and collagen type I content [4].

This concept justifies the resistance of bone to sudden loads/impacts like jumps or falls: if there is some degree of deformation, there is later a return to the original form.

However, bone tissue is not always capable of withstanding all the loads that are applied to it, and this failure occurs when the imposed force exceeds the elastic deformation capacity, which may cause a complete or incomplete fracture. In this last situation, permanent deformation occurs in the tissue even after the load is removed, triggering microfractures and trabecular disruption and preceding macroscopic rupture [3].

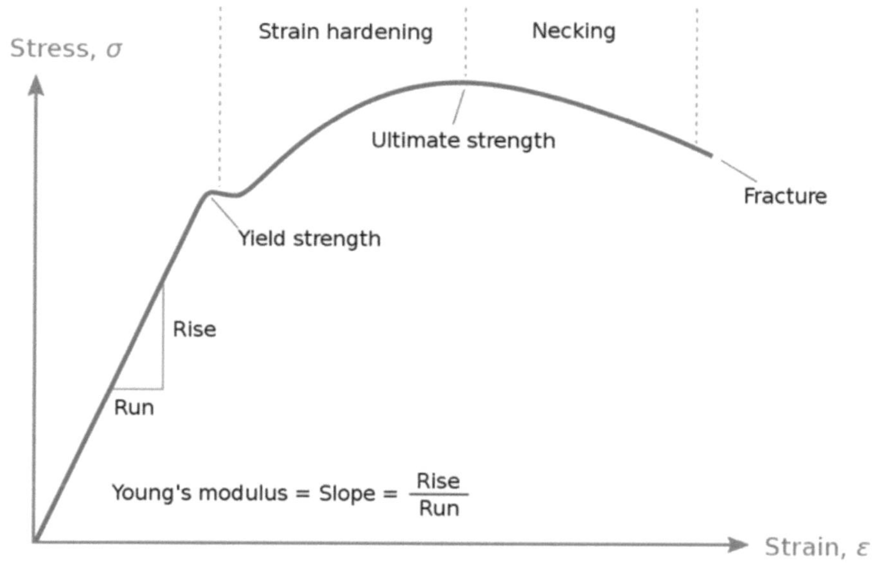

Figure 1.
Graphic representation of the stress/strain curve, and the biomechanical concepts of young modulus, yield, and fracture point.

When a supraphysiological force is applied, the bone will be deformed. The mechanical properties of bone are well represented by the load/stress-strain curve (**Figure 1**). The load/stress response of the bone directly depends on the length, thickness, density, shape, and type of bone (cortical vs. cancellous), so the stress-strain curve represents the structural properties of the object. Bone is characterized by a stress-strain curve with initially a linear response, the so-called elastic deformation. From the point at which the response ceases to be linear to the load and starts to express a curve (physiological limit point), plastic deformation appears and microfractures occur at the structural level, which can culminate in macroscopic fractures if the load is not interrupted or maximum force is reached (fracture point) [3].

2.3 Strength and stiffness

Strength denotes the ultimate load a material can withstand before a catastrophic failure, which is also designated as the fracture point (**Figure 1**) when this concept is applied to bone [1–3].

The stiffness of a biomaterial or bone tissue is the mechanical property that characterizes and quantifies the changes in the original shape when a force vector or a load is applied to it. A graphic that represents stiffness is called a load/displacement curve, and the relationship between stress and strain for materials, or load and displacement for structures, can help us understand these properties better. The slope of the straight part of the curve that ascends represents the elastic modulus or stiffness. The steeper the slope of this part of the curve, the stiffer the material. The strain (or change in shape) in this part of the curve is elastic, which means the material can return to its original shape after the force is removed. There is a point on the curve, called Y, known as the yield point or yield load where the curve stops being nonlinear (**Figure 1**). At this point, the strain exceeds the material's ability to recover from its original shape, and the material gets a permanent change in shape if the load is removed. This point shows where the material changes from elastic to plastic deformation. This point is important for clinical reasons because it means that the bone acquires a different shape from its original [2].

This permanent change in shape, called plastic (instead of elastic) deformation, occurs when covalent bonds break at a molecular level. The point on the curve where the material breaks or fails is called U, or ultimate failure/fracture point (**Figure 1**). The curve also shows how much energy the material can absorb during the loading process [2]. This is called toughness, and it is represented by the area under the curve [2].

2.4 Elastic modulus of young

The elastic modulus, also called the modulus of Young (Y), represents the relationship between stress and strain and is one of the most useful parameters for mechanically comparing biomaterials. It is calculated by applying the formula represented by Eq. (3), where the slope of the stress is plotted versus the strain curve (**Figure 1**). It has the same units as stress (N/mm^2) because strain does not have any units [3]. It quantifies the relationship between tensile/compressive stress. Young's moduli values are normally so large that this parameter must be expressed in gigapascals (GPa) instead of in Pascals. The components of the formula for Young's modulus calculation are: σ (force per unit area) and axial strain, ε (proportional deformation) in the linear elastic region of a material and is determined using the formula (Eq. (3)):

Young modulus formula:

$$E = \frac{\sigma}{\varepsilon} \tag{3}$$

The elastic modulus of Young is a measure of the linear elasticity of a material. This parameter allows grading materials in two categories: flexible and rigid. Materials with higher Y values are considered rigid. The bone tissue, for example, is included in the rigid category with a value of 15 GPa but is less rigid when compared to materials used in the manufacture of orthopedic implants. The elastic modulus of stainless-steel implants is usually 188 GPa, pure titanium is 116 GPa, and titanium alloy (Ti-6Al-4 V) is 113 GPa. A single value of Y assumes a linear relationship, which is true for metals (until their yield point) [3].

2.5 Area moment of inertia (AMI)

The area moment of inertia is a geometric parameter to be considered when the mechanics of implants are studied. The AMI is a measure of the resistance of materials exclusively related to flexion loads. This parameter is only influenced by the geometry and not by the composition of materials. Implants manufactured with bigger AMI have the least probability of structural collapse when submitted to higher flexion loads (higher stiffness to flexion loads). AMI does not take into account material properties, and for that reason, AMI must be only used to compare different constructs of the same material. AMI is a geometric parameter that is calculated based on the dimensions of the structure in the direction of bending. For a circular implant (e.g., a pin or interlocking nail), the direction is not relevant, and the formula used for this particular type of implant is the following (Eq. (4)):

AMI formula for circular implant:

$$\frac{1}{4.p.r4} \tag{4}$$

In Eq. (4), the radius is raised to the fourth power, so a small increase in the diameter of a pin or other circular implant has a large impact on its bending stiffness.

This concept can be illustrated by the following example: if the AMI of a 2.4 mm pin (3/32 inch) is 1.6 mm^4 and for a 3.2-mm (1/8 inch) pin is 5.1 mm^4, an increase of 33.3% of the pin diameter results in an increased AMI value, 3 times larger than an original pin.

For solid rectangular structures, AMI is calculated using a different formula (Eq. (5)):

AMI formula for solid rectangular structures:

$$\frac{1}{3.b.h3} \tag{5}$$

In the AMI formula, *b* is the width, and h is the height in the direction of bending. Plate thickness is an important parameter because this dimension is cubed. However, for bone plates, the presence of the screw holes adds complexity to this calculus. At the screw holes, the AMI is usually less than half the value that would be calculated from its external dimensions [5].

2.5.1 The impact of plate orientation on AMI

If the direction of the bending force of a fracture is known or the vector force can be simplified to a craniocaudal direction, the orthopedic surgeon can also use this concept of AMI to consider alternative plate locations. The classic example of this concept is for distal radius fractures. In this type of fractures, the primary direction of bending is considered to be in the craniocaudal plane; if a 2.7-mm LC-DCP (Limited Contact-Dynamic Compression Plate) is placed on the medial aspect, a higher AMI value (solid section of approximately 111 mm^4) will be measured when compared to a 3.5 mm LC-DCP placed on the cranial aspect (AMI of 30 mm^4), because the height of the 2.7 mm plate in the direction of bending is 8 mm (almost 3 times greater), compared with 3.3 mm for the 3.5 mm plate [2].

Another variant that influences the AMI is the position of the implant regarding the neutral axis of the bone, which is represented by the medullary canal of the bone. If the implant is positioned more distant from the neutral axis, the implant is less efficient to resist the bending forces. For the mentioned reason, the interlocking nail is the most mechanically favored implant to resist bending forces [3].

The use of AMI helps the decision-making process for choosing the osteosynthesis method but is not exclusively based on this parameter. Every long bone has a tension side and a compression side when axial loading is applied to the bone that will cause deformation, promoting bending. When the axial loading is applied and the bone bends, one side will experience tension and the cortices suffer traction. At the same time, the opposite side of the bone and the bone cortices will experience compression. Every long bone has a neutral axis that corresponds to the medullary cavity and that does not suffer compression or traction forces, and also has a tension and compression sides [1–3].

By using the AMI and tension/compression sides defined for each bone, in the decision-making process of orthopedic surgery, two premises must be fulfilled to succeed:

1. Positioning the plate on the tension side of the bone, because this side will be more mechanically challenged by the traction forces. When bending forces are applied to the bone, the tension side will suffer distraction and the plate protects the bone. On the opposite side, the bone suffers compression, which is less mechanically challenged;

2. Fractures not anatomically reconstructed in the compression side (gap fracture) due to a comminuted fracture or by losing far cortical support will suffer an inversion of the former dynamic. During the flexion loading of the bone, the plate side will suffer compression, which predisposes to plate failure. Clinically, in these cases, we must augment AMI by increasing the plate thickness, increasing the working length of the plate, combining implants (orthogonal and bilateral plating and/or intramedullary pinning), or using buttress plates without screw holes in the working length of the plate (biological osteosynthesis plate) [3].

2.6 Working length of plate

In locking plates, the distance between the proximal and distal screw in closest proximity to the fracture is defined as the "working length" of the plate (**Figure 2**). If a plate is compressed against the bone (dynamic compression plate), the working length is the distance between the bone ends of the fracture span by the plate (**Figure 2**). A correlation between plate working length and stiffness of the construct, plate strain, and cyclic fatigue properties of the plate has been shown [5, 7–10]. Another fundamental aspect in high-strain fracture management (e.g., simple transverse fractures) is load sharing between the stabilized bone and the plate; not addressing this aspect in osteosynthesis increases the risk of cyclic fatigue and early failure of the plate.

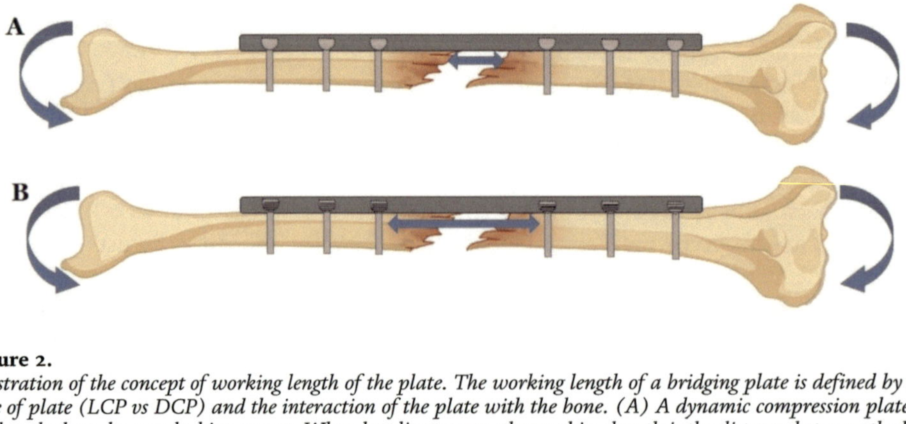

Figure 2.
Illustration of the concept of working length of the plate. The working length of a bridging plate is defined by the type of plate (LCP vs DCP) and the interaction of the plate with the bone. (A) A dynamic compression plate is held to the bone by non-locking screws. When bending occurs, the working length is the distance between the bone ends of the fracture span by the plate. (B) The locking plate spans the same gap, but because the plate is not in direct contact with the bone, the distance between the nearest proximal and distal screw to the fracture line is defined as the working length.

Also, the plate length influences the screw loading. For a given amount of bending moment, a longer plate produces markedly less pull-out force than a short plate due to an improvement of the working leverage for the screws.

The effective application of plate length is another concept. The farthest screws determine the effective usage of plate length and contribute to fracture gap stability. A long plate produces markedly less pull-out force than a short plate.

2.7 Improved anchorage by diverging screws

Locked or non-locked screws with divergent inclinations improve the anchorage considering that a bigger amount of bone is displaced when compared to parallel screws when an equal pull-out load is applied.

2.8 The helicopter effect

The tightening of the first screw (head locking) in one extreme of the plate without stabilization of the other end of the plate will cause the "helicopter effect". This effect also occurs if only two screws were applied in auxiliary plating used in the orthogonal plating technique. In orthogonal plating, the auxiliary plate must have a minimum of two screws applied for the segment to prevent the helicopter effect (**Figure 3**).

2.9 The strain theory

The control of interfragmentary micromotion is the key point for correct fracture stabilization. The knowledge of the factors that dynamically influence the distances between fracture fragments is fundamental to controlling micromotion. The strain theory is the most important concept used in the decision-making process, regarding fracture osteosynthesis from a mechanical perspective.

Essentially, strain or relative deformation is the amount of movement (distancing/approaching) between fracture fragments relative to the original distance (gap).

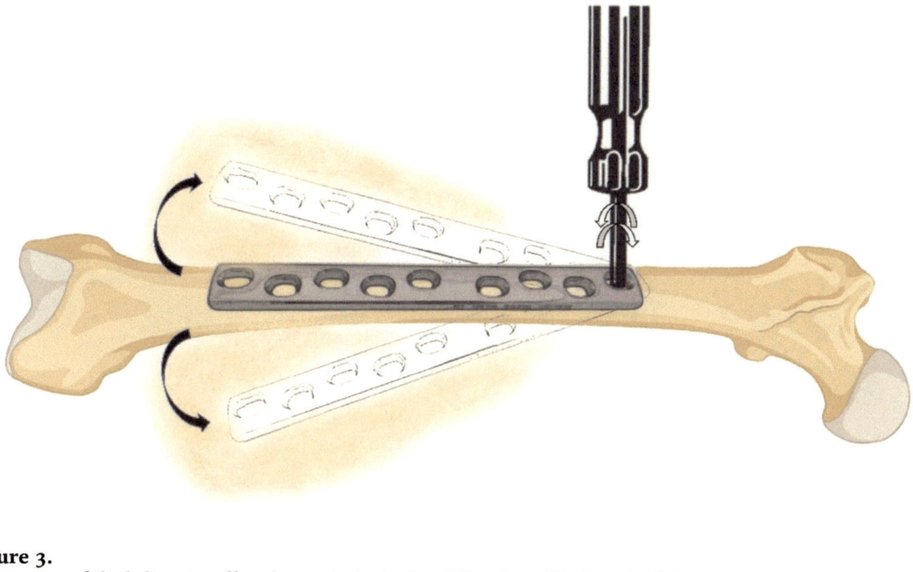

Figure 3.
Illustration of the helicopter effect due to the lack of stabilization of both ends of the plate.

It is expressed as a percentage of movement, that is, movement of gap/original fracture gap when the fragments were subject to mechanical stimulus (weight support, muscular contraction, and passive movement, among others). The calculus of interfragmentary movement in a laboratory environment is obviously more precise than in clinical settings. In clinical scenarios, several factors can influence the magnitude and direction of the fragments in the process of distancing/approaching movement; among these factors are the great variability of fracture patterns and the correspondent mix pattern of strain simultaneously present at the fracture lines [3].

Mathematically, the strain is determined by the formula (Eq. (4)):

Formula for strain determination:

$$E = \frac{\Delta L}{L}$$

where the E is the strain expressed by % value, ΔL is the variation of interfragmentary space (gap variation), and L is the initial gap. By the equation, is possible to infer that if the initial gap is bigger, the final strain value inversely will be

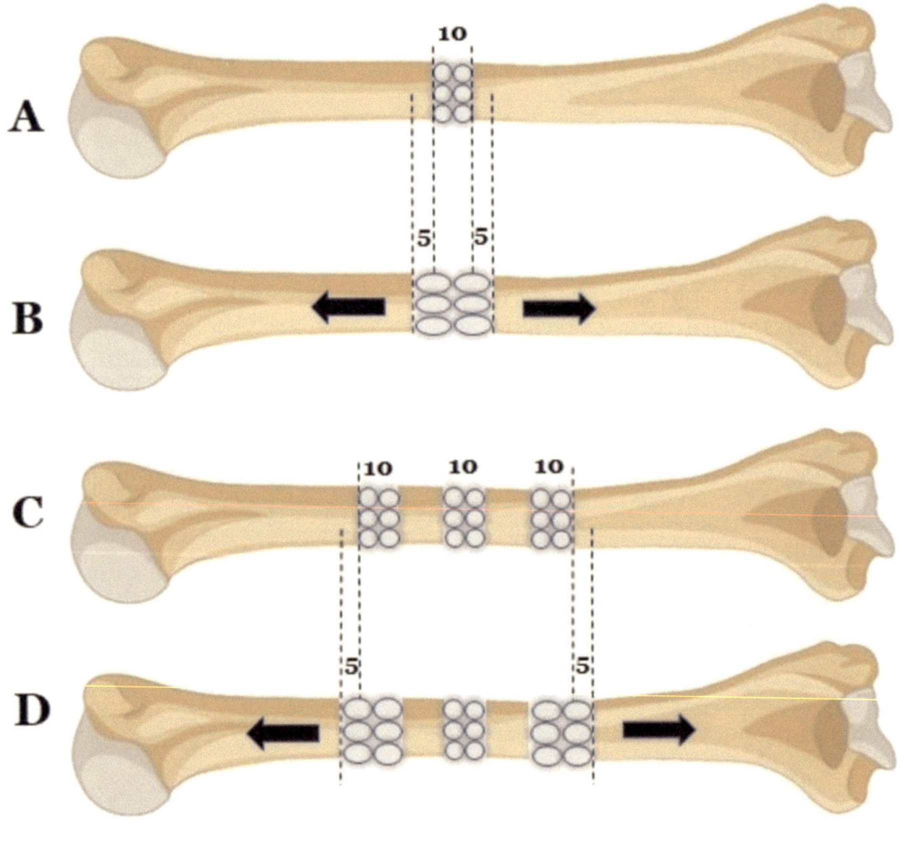

Figure 4.
Illustration of interfragmentary displacement. A and B—simple line fracture with a gap of 10 mm; when the displacement of 5 mm between both fragments occurs, a strain of 100% is produced; C and D—multiple line fracture with a gap between fragments of 10 mm, totalizing a fracture gap of 30 mm; when the same displacement of 5 mm in each major fragment takes place, a total of 10 mm is also added to the fracture gap; however, in this case, the final gap undergoes deformation (strain) of approximately 30%, because the final displacement was distributed over all fragments.

smaller and when a force is applied to a fracture site if the variation of the gap is bigger, the strain also increases.

The illustrative example of two different interfragmentary displacement scenarios is given in **Figure 4**. In the first two images, a simple line fracture with a 10 mm gap between bone fragments is shown. When both fragments are displaced by 5 mm, a displacement (strain) of 100% occurs. In the lower images, a multiline or communitive fracture is depicted, and there is a gap of 10 mm between each fragment, totaling a gap of 30 mm. When the same displacement of 5 mm in each major fragment takes place, a total of 10 mm is also added to the fracture gap; however, in this case, the final gap undergoes deformation (strain) of approximately 30%, because the final displacement was distributed over all fragments.

In the biological context, the strain concept is used to explain the relative deformation and its effects on bone tissue regeneration. In bone callus formation, the tissue can resist a different amplitude of elongation (distancing of fragments). If the movement exceeds the critical value of elongation, there will be dysfunction at a cellular level and, consequently, no delay in the tissue formation. In bone regeneration, the predominant cells in each phase show different tolerance to different magnitudes of elongation movements. As bone regeneration progresses, the tissue is less tolerant, demanding a more rigid mechanical environment (with less micromotion). During the inflammatory phase, the granulation tissue is the most tolerant to movement when compared to cartilaginous or bone tissue in subsequent phases (**Table 1**).

To illustrate the difference in instability tolerance between a simple fracture and a multi-fragmentary fracture, consider the following scenario:

Assuming both fractures have the same initial gap width (5 mm) and overall displacement (5 mm) in (A) and (B), the full displacement (5 mm) is active within a single gap in a simple transverse fracture (A), resulting in a strain of 100%, which is the limit of tolerated strain for granulation tissue. In contrast, in a multi-fragmentary fracture with five gaps (B), the overall displacement is shared among the gaps, resulting in each gap displacing from 5 mm to 6 mm, and the resulting strain is only 20% [3].

Additionally, and from a mechanical point of view, different fracture patterns presented different strain behaviors when subjected to the same stress load. Generally, the larger is the lever arm, the more movement at the interfragmentary interface will be observed; this occurs in single-line fractures (transverse and oblique) in which the fragments are long relative to the fractured section. In the previous scenario, the gap is small and the variation is large, determining a high-strain fracture environment. In simple words, these fracture lines are more sensitive to load/movement forces. In the opposite scenario, multiline fractures, the lever arm is smaller (multiple fragments smaller in length) and the total interfragmentary space (gap) is inevitably larger; with greater gap and equal length variation, with the same stress/load forces applied, the strain value will be smaller (Eq. (4)). For this reason, these fractures are considered low strain or less sensitive to movement or load forces [3].

Tissue type	Tolerance to elongation (%)	Tolerance to shortening (°)
Granulation tissue	100	40
Cartilage	15	5
Bone	2	0.5

Table 1.
Tolerance of tissues of the osteogenic pathway to elongation and shortening.

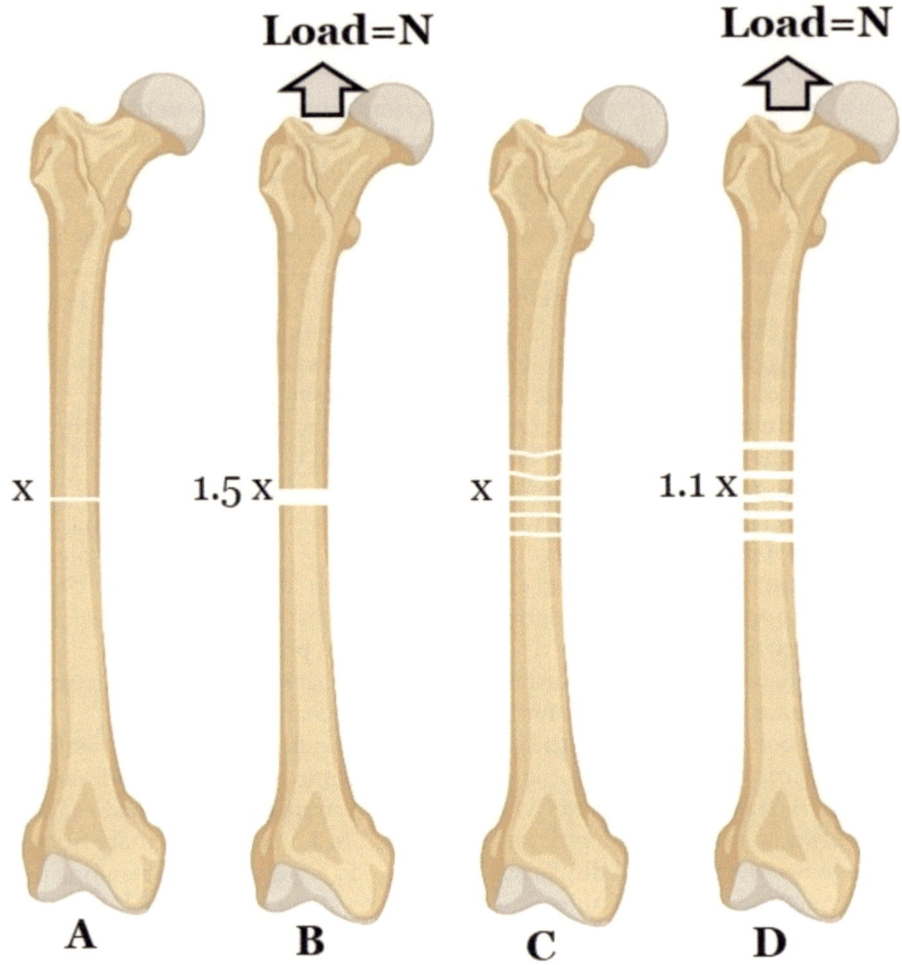

Figure 5.
Illustration of fracture lines with high and low strain. High-strain fracture pattern, A and B—fracture line without load and with a little gap (x); when the bone is loaded, the fracture gap increase by 50% (1.5 x); low-strain fracture, C and D—multiple line fracture with a gap equivalent to fracture line in A; however, when the bone is loaded, the displacement is more subtle among fragments (1.1 x) because the displacement is distributed between all fracture fragments.

Clinically, for the surgical decision-making process, the surgeon should consider two groups of fractures based on the strain theory:

1. High-strain fractures (simple fractures with long fragments) are more sensitive to motion, and interfragmentary movements can easily delate the process of bone regeneration (**Figure 5**). Knowing that, we must choose a method of osteosynthesis which leads to anatomical reconstruction with interfragmentary compression and rigid fixation (reducing the gap and ΔL).

2. Low-strain fractures (multiple line fractures or complex fractures with smaller and short fragments) are more tolerant to movement, allowing more load without major consequences (least chances to overcome tissue tolerance) (**Figure 5**). In these cases, a more elastic and less invasive approach must be

chosen over osteosynthesis methods. This type of fracture is caused by high-energy trauma with more need to preserve the soft tissue envelope of the fracture. Considering this scenario, bone osteosynthesis methods rely on bridging plates (biological osteosynthesis plates) and external fixation.

It is important to remind that several biological and clinical factors influence bone regeneration such as age, time elapsed since trauma, trauma intensity, and soft tissue disruption (open or closed fracture). For this reason, the decision process in bone osteosynthesis is multifactorial and should not be based exclusively on mechanical factors like the strain theory.

2.10 Wolff's law

Bone is a dynamic tissue, and the response to internal and external mechanical stimuli can determine bone density and the organization of bone trabeculae and correlate with the magnitude and direction of compressive and tensile stresses of loading. In the late nineteenth century (1892), Wolff's law was proposed by Julius Wolff, a German anatomist and surgeon, as a mathematical law that described the response of bone to mechanical loading. This law described the functional adaptation of bone to mechanical loading and is supported by several experimental and comparative studies over time. Increases in the loading of bone tissue are known to generate the formation of new bone tissue, which increases mechanical rigidity [6]. Similarly, decreases in mechanical loading, particularly associated with prolonged non-weight-bearing lameness, lead to adaptive resorption or osteopenia of bone tissue conducting to a decrease in mechanical rigidity [6]. One of the classic examples of Wolff's law is the femoral trochlear groove formation by the pressure that the patella exerts on the bone. In dogs with congenital medial patellar luxation, with the lack of pressure by the abnormally positioned patella, the trochlea can be shallow or absent. Another practical example of this law is that bone is generally stronger and stiffer in the direction in which the greatest loads are most commonly imposed (e.g., the long axis of the femur).

2.11 The piezoelectric effect

The piezoelectric effect is a physiological characteristic of certain materials that generates an electric charge in response to a supported mechanical load. The suffix Piezo is derived from the Greek *piezein*, which means to press [7].

Bone has piezoelectric properties because of the highly oriented and patterned structure of collagen type I, and collagen's ability to respond to mechanical loads [7, 8]. When a shearing force is applied to collagen fibers, and the bundles glide past each other, an electric charge is generated. Collagen also has significantly lower elastic moduli than the bone's corresponding mineral component, which makes collagen microfibrils experience the greatest load when strained. Experiencing the greatest load under force deforms collagen fibbers, and this deformation leads to the piezoelectric effect [8]. The role of collagen's piezoelectricity in bone regeneration and remodeling is related to the formation of electric dipoles that stimulates the osteoblasts to promote mineral deposition in the extracellular matrix, increasing bone density. Clinically, when a fracture occurs, the collagen's piezoelectricity is potentiated with an additional mechanism for osteocytes to sense areas with more stress; the generated piezoelectric charge would be greater in stressed areas, which is produced when the bone suffers deformation and negative charges are produced on the tension side and positive charge on traction side, generating bone growing by electric current [7].

3. Biomechanics of bone fracture

Bones from the appendicular skeleton are continuously subject to physiologic and non-physiologic mechanical forces. Physiologic forces are generated through weight-bearing and muscular contraction during physical activity or even at rest. Didactically, it is established that force vectors act isolated in long bones such as flexion or bending, axial compression, tension, and shear and torsion forces; however, clinically, one force vector is predominant. In healthy animals, the physiologic loads applied to bones rarely exceed the yield point, or more practically, physiological forces do not cause plastic (permanent) deformation of bone (**Figure 6**). Nonetheless, when non-physiologic forces are the result of externally applied loads (vehicular trauma, horse kick, fall from height, and gunshot), it is easily exceeding the yield point and load-bearing capacity of the bone is easily exceeding, and as consequence, fracture will occur (**Figure 6**). The resistance of the bone will vary according to the direction of the load, and depending on the intensity and type of forces applied to the bones, different fracture lines will form. In general, oblique fractures originate from supraphysiological axial compression forces; transverse fractures are related to tension (avulsion) and flexion forces applied on opposite sides of the long bone; spiral fracture lines are expected from torsional forces, which create an angular line running around the circumference of the bone and a longitudinal one joining the two ends of the spiral. The combination of forces will give rise to other patterns of fracture lines, such as segmental fractures (butterfly fragment) caused by shear failure on the compression side before the tension fault line; it then propagates throughout the bone, creating the compression side plus a fracture line, usually a single line fragment. The combination of flexion and compression forces potentially generates early failure on the compression side, and a larger fragment will break loose (major butterfly fracture).

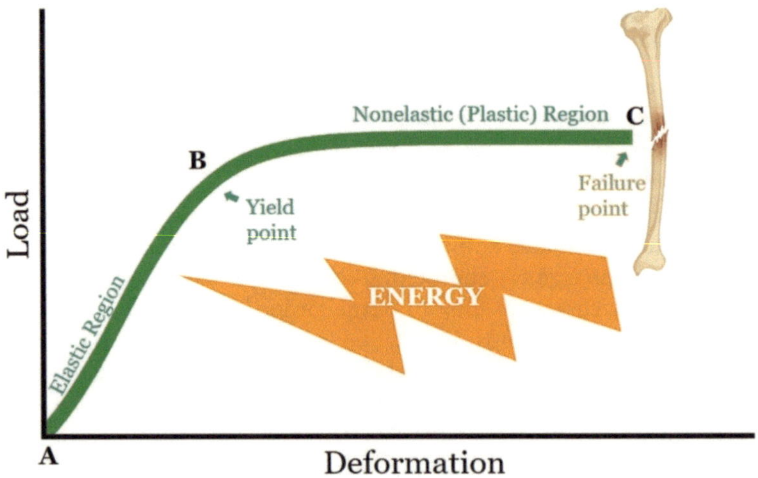

Figure 6.
Load/deformation curve of long bones (A—starting load, B—yield point of deformation, C—fracture or failure point).

The minority of clinically encountered fractures are a result of pure isolated loading; the majority of fractures presented to a veterinary surgeon are caused by more complex loading situations. Most clinical fractures are produced by a combination of three or more loading modes, resulting in a fracture line initiation and progressing in numerous planes. For example, a fracture caused by vehicular trauma is often the result of a combination of bending, shear, and torsional loads. Additionally, the values of the different loads would most likely be different, causing further variations in the fracture patterns observed.

Fractures caused by high-energy trauma (e.g., road traffic accidents) involve the greater accumulation of energy associated with the combination of forces, often resulting in greater fragmentation (comminuted lines). These fractures will also cause greater muscle damage and vascular compromise. The direction in which the energy is applied to the bone is as important as the intensity with which it propagates. Energy is absorbed by the bone and then released with the fracture. Damage applied to soft tissue and bone is proportional to the amount of energy released, and it is concluded that complex fractures are associated with greater soft tissue enveloping lesions.

3.1 Forces acting on bone fracture

3.1.1 Bending

The definition of bending comprehends the axial compressive load that is applied eccentrically (off-center) to the bone column. Studies have demonstrated that 85–89% of the predominant normal physiological stresses that most bones experience during weight bearing are from bending loads due to the curvilinear shape of the femur, humerus, and, in some breeds, tibia e radius [1]. In these cases, the axial compressive load is applied eccentrically during locomotion [1]. The convex face of the bones experiences the maximal load of tension forces, whereas the concave side experiences the maximal load of compression forces. The bone column experiences a gradient distribution of this opposite force perpendicular to the bone, the axis being the center

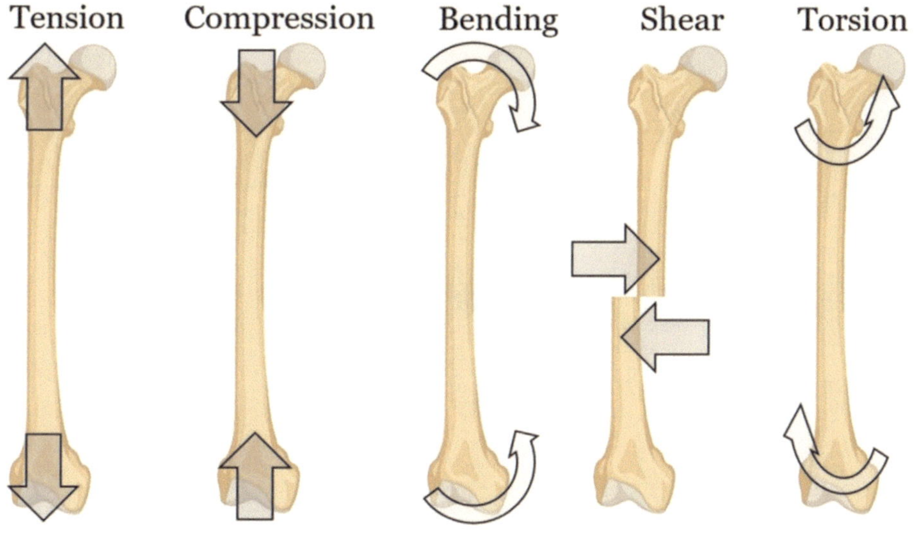

Figure 7.
Different force vectors acting on long bone fractures.

of the medullary cavity, an imaginary point where there are no tensile or compressive stresses and strains. This plane is called the neutral axis or neutral plane of the bone and experiences no axial stress (**Figure** 7) [1]. The magnitude of the compressive and tension forces acting on the bone increases as the distance from the neutral axis increases (**Figure** 7). When a supraphysiological bending load exceeds the yield point and the load-bearing capacity of bone, for example, in extremely soft bone such as immature or diseased bone, when it is subjected to a bending load. A fracture line starts at the tension side (because the cortical bone is weaker in tension than in compression) and propagates to the compression surface, producing, in most cases, a transverse fracture.

If an internally generated shear stress is added to these forces, it results in a short oblique fracture line toward the compression band surface of the bone. Two oblique fracture lines can occur near the compression surface, if the magnitude of shearing forces increases, forming a loose wedge. This fracture pattern is referred to as a butterfly fracture and is a result of two divergent planes of shear stresses near the compression surface.

3.1.2 Compression (axial)

Compression loading, also called axial compression, is produced when equal and opposite loads are applied toward the center and parallel to the longitudinal axis of the bone, causing compressive stress and strain within the bone (**Figure 8**). In long bones, compressive loads cause a decrease in height and an increase in width. Maximum compressive loads occur on a plane perpendicular to the applied load and can be defined by a series of small forces directed toward the center of the bone that

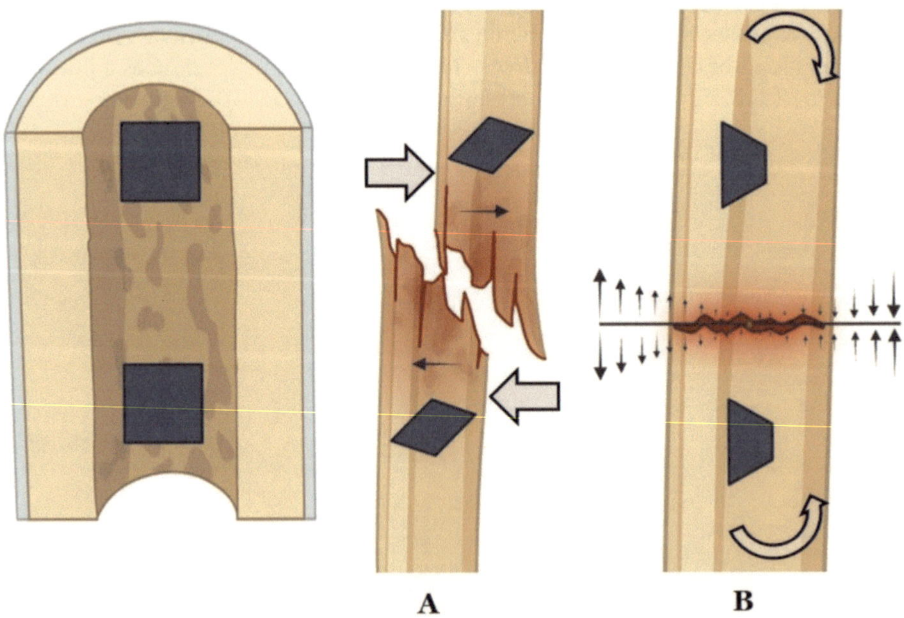

Figure 8.
Illustration of shear and bending forces acting on a long bone, A—shear loading causes angular deformations, B—bending loading induces tensile loading along the convex surface and compressive loading along the concave surface causing a transverse fracture pattern.

potentially can compact or crush the bone. Rationally, we would expect that compression fractures developed perpendicularly to the applied compressive load will crush the bone. Nonetheless, the perpendicular tensile strain is usually not too important, because the expansion of cortical bone is highly unlikely and internally generated tensile strain also develops outward from the center of the bone, perpendicular to its longitudinal axis.

Nonetheless, compression loading also produces internal shear loading that develops oblique to the longitudinal axis and is maximal on a plane of 45° from the axis of compressive loading (**Figure 9**) [9]. Macroscopically, the fracture line of bone loaded under pure compression is typically a short oblique fracture and is created by these internal shear stresses, generated partly because of the bone's anisotropy and the fact that bone is weaker in shear forces and more tolerant in compression loads. These oblique fracture configurations produced by compressive loading are commonly seen clinically with jump or fall injuries of the distal tibia and radius (bones that are loaded along their central axis) [9].

A transverse fracture pattern also can appear as a result of compressive loading and is occasionally seen in vertebral bodies or the growth plates of long bones in young animals, also called impaction or impacted fracture (type V or VI Salter-Harris fracture) [1].

3.1.3 Shearing

Shear loads occur when a force is applied parallel to the bone's surface, causing it to have a tendency to slide past another surface and causing an angular deformation (**Figure 9**). With the shear forces acting in opposite directions on opposing surfaces, shear loads within the bone lead to deforming it in an angular manner (right angles within the bone are deformed to acute or obtuse angles). In general, the bone offers the weakest strength when subjected to shear forces. Therefore, bone fractures along the plane of maximal shear stress. Clinically, fractures developing from shear loading often occur in the metaphyseal region of long bones with high cancellous bone content [9].

A classic example of fracture that occurs in small animals as a result of pure shear loading is the fracture of the lateral aspect of the distal humeral condyle seen in immature animals (Salter-Harris type IV). This fracture occurs as axial compressive forces are transmitted from the foot through the head of the radius to the lateral and/or intercondylar component of the condyle of the distal humerus, resulting in a concentration of shearing forces at these regions of the distal humerus, producing a classical type IV Salter-Harris fracture [1]. Other common fractures created by shear loading would include "T" or "Y" intercondylar fractures of the distal humerus, fractures of the tibial plateau, isolated condylar or intercondylar femoral fractures, fractures of the glenoid cavity of the scapula, vertebral body fractures, and carpal or tarsal bone fractures. As previously described, shear loads also occur in most long bones subjected to pure axial compression, resulting in short oblique fractures along the plane of maximal shear stress [1, 9].

3.1.4 Stress concentration or stress risers

Osteopenia and bone defects on bone structure caused by iatrogenic conditions such as drilled holes (biopsy tract, bone graft collection, or screw removal) or acquired conditions like neoplasia, bone cysts, and bone infection (bacterial or fungal

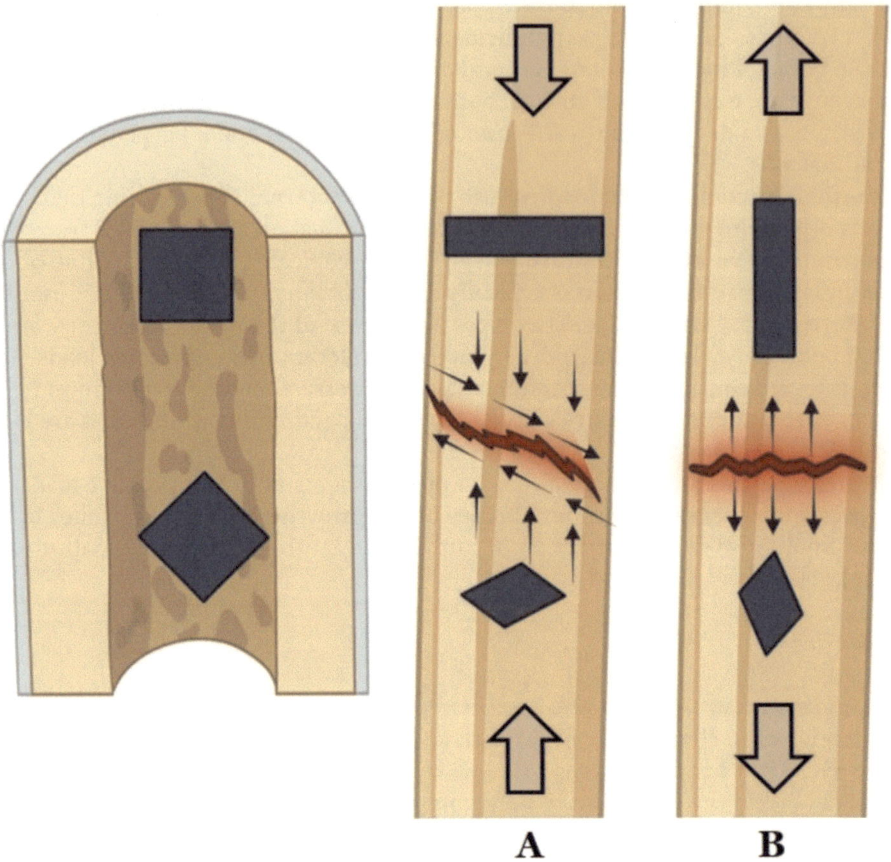

Figure 9.
Illustration of stress and strain produced by compressive and tensile loading. A—Compressive loading induces compressive and shear stresses and strains that, if excessive, may induce a short oblique fracture. B—Tensile loading induces tensile stresses and strains, which, if excessive, induce a transverse fracture.

osteomyelitis) cause stress concentrations in bone that can initiate failures [10–12]. These areas of stress concentration can lead to local stress risers in the bone near the defect, which is many times higher than the stress physiologically applied to the bone. The concept of the stress concentration effect is based on the mechanical phenomenon that physiological loads must flow through the bone and, in a healthy bone tissue, without defects or heterogeneity. The applied forces flow equally through all regions, creating equal stress throughout. However, in bone with defects (e.g., holes from removed screws), the load cannot flow through the areas with defects and thus must flow around the holes. This leads to a concentration of stress in points adjacent to defects and osteopenia areas [9]. The clinical consequence is that bone stress-rising points break at lower loads than homogenous bones. The weakening effect of a stress riser is particularly noted for torsional loading where the decrease in strength may approach 90% and is proportional to the defect size [10]. However, defects smaller than 10% of the bone diameter may be of negligible significance in torsional resistance and may resist under physiological loadings [13].

Another form of stress concentration is illustrated by the difference in the elastic moduli (stiffness) of two materials (e.g., stainless steel and bone) under load. The stress concentration in this example is due to the evidence that the modulus of a material determines its response to an applied force: with high moduli materials, the strain or deformation is inferior to low moduli materials for equal load. As a consequence of that difference in modulus values, the flow of homogenous force is interrupted causing stress concentration. Frequently, clinical cases are reported in bone areas at the limits of the joint prosthesis or stainless-steel bone plating. As the materials are loaded, the bone exhibits greater elastic deformation, creating shear stress at the bone-implant interface [9].

3.1.5 Tension

Tensile loading of bone results when equal and opposite loads are applied away from each other outward from the bone's surface and along its longitudinal axis, as a result of supraphysiological stress, causing the fracture line to be orientated on a plane perpendicular to the axis of loading (**Figure 9**). This mode of loading is primarily due to the contraction of muscles or the effects of ligaments and tendons at bone prominences such as tuberosities, tubercles, and trochanters, where a pure tensile loading is exerted over their cross-sectional area. Clinically, fractures with transverse patterns perpendicular to the applied load are predictably produced and often seen at traction of apophyses such as the olecranon process, tuber calcaneus, and tibial tuberosity (**Figure 9**). Fractures of the patella and avulsion fractures of ligamentous insertion are also exampling where tensile forces predominate and cause a transverse fracture. Because cancellous bone is much weaker under tension than cortical bone, fractures occurring due to tensile loading often occur in regions that have more cancellous than cortical bone, such as bone prominences.

3.1.6 Torsion

When a torsional load is applied to a long bone in such a manner that causes it to twist about an axis (usually the long axis of the bone), that results in the generation of shear, tensile, and compressive forces (**Figure 10**). Specifically, the torsion force causes a shear stress that is distributed throughout the bone. As in bending, there is a gradient in the magnitude of loading, proportional to their distance from the central (long) or neutral axis. Under torsional loading, maximal shear stresses are produced on planes perpendicular and parallel to the central axis. Tensile and compressive stresses are distributed perpendicular to each other and on a diagonal plane to the neutral axis (**Figure 10**). The fracture begins along a plane of maximal shear stress orientated parallel to the neutral axis. The fracture then propagates along the plane of maximal tensile stress creating the typical spiral fracture configuration. Clinically, spiral fractures are commonly seen in the narrow diameters of the distal tibial and distal humeral diaphysis, where the area moment of inertia is relatively small (thus the resultant shear strain from torsional stress is relatively high).

4. Biomechanics of bone tissue

The fracture behavior of bone is influenced by its viscoelastic, anisotropic, and heterogeneous mechanical properties. The stress-strain behavior of bone is dependent on the rate of loading, which is characteristic of a viscoelastic material [3, 14]. If the bone is loaded at a high rate, such as occurs with vehicular trauma or gunshot injury,

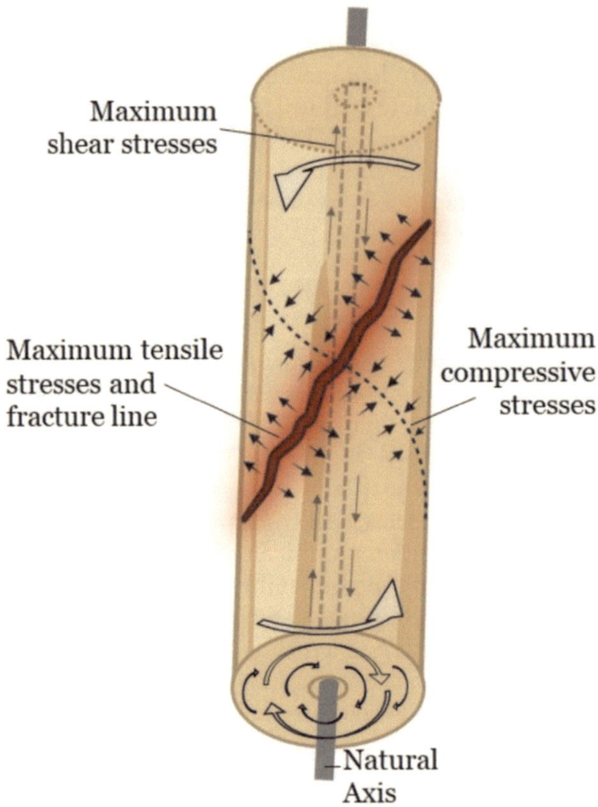

Figure 10.
Illustration of the shear, tensile, and compressive stresses and strains at supraphysiological loads causing a spiral fracture pattern.

its stiffness (Young's modulus), ultimate strain, and energy-to-failure increase. The clinical significance of the high toughness of healthy bone is that if a high-rate loading causes macroscopic failure or fracture, as opposed to just distributed microscopic interfacial failures, the large release of the absorbed energy will cause marked comminution and injury to surrounding soft tissues [15]. Bone is considered a material with anisotropic properties; as a consequence, the values of strength and stiffness are a function of the direction of applied loads regarding bone structure (**Figure 11**) [3, 14].

4.1 Cortical vs. cancellous bone material properties

All bones are composed of a combination of cortical (compact) and cancellous (trabecular) bone. Both cortical and cancellous bones are formed from an inorganic mineralized matrix called hydroxyapatite, which is primarily calcium and phosphate. Hydroxyapatite (HA) is a naturally occurring calcium phosphate mineral characterized by the chemical formula $Ca_{10}(PO_4)6(OH)_2$. HA-like compounds compose approximately 60–65% of bone's dry weight [16]. The inorganic matrix is combined with an organic nonmineralized matrix (35–40% of bone's dry weight) [16]. By contrast, the organic extracellular is significantly more complex and consists mainly of

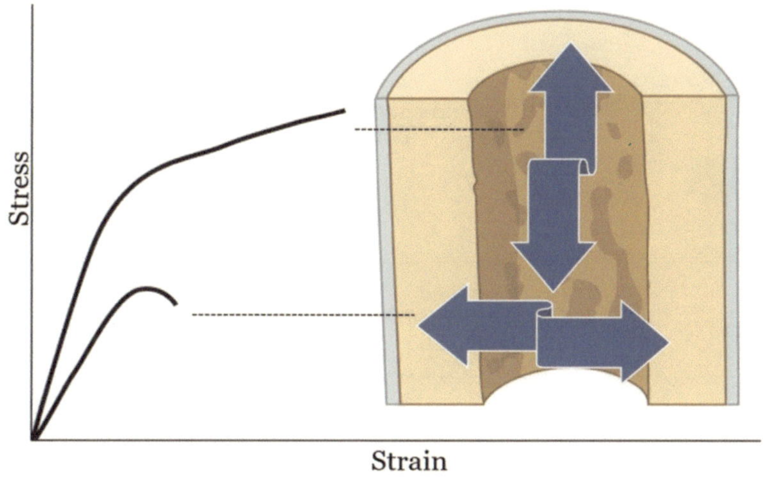

Figure 11.
Stress-strain curve depicting the anisotropic behavior of bone. Load forces of tension were applied in two different orientations: parallel and perpendicular to the longitudinal axis.

collagen type I (90%) and noncollagenous proteins (10%) such as glycosaminoglycans, water, and cellular elements [16]. The inorganic matrix imparts strength and rigidity to the bone, and the organic matrix gives it flexibility and resiliency [1].

The cortical bone always surrounds the cancellous bone; however, the relative quantity of each type varies from one bone to another as well as according to the specific location within a particular bone (diaphysis vs. metaphysis or epiphysis); cortical bone is designed to give strength and stiffness to the bone [3]. From a mechanical standpoint, cancellous bone is designed to absorb a tremendous amount of energy and transmit load [1].

Both cortical and cancellous bones have inorganic and organic components; however, one of the primary differences between both bone types is the different percentages of organic versus inorganic matrix of each type. Structurally, this difference influences the porosity and apparent density and consequently the mechanical behavior of each type of bone when submitted to loads.

Porosity is defined as the volume of bone occupied by nonmineralized tissue. Cortical bone is composed primarily of inorganic mineralized matrix and therefore has low porosity. The porosity of cortical bone has been estimated to vary from 5% to 30% and in the cancellous bone, it can vary from as little as 30% to as much as 90% [17].

Apparent density is a measurement related to porosity and is directly related to its inorganic mineral content, being the mass of the bone tissue divided by the bulk unit volume of bone tissue, including mineralized bone and marrow space [17]. Cortical bone typically has a higher apparent density than cancellous bone tissue [17].

The differences in porosity, or apparent density, between cancellous and cortical bone dramatically affect their behavior when the two types of bone are submitted to loads (**Figure 12**). Cancellous bone initially exhibits elastic behavior followed by a yield, which occurs as bone trabeculae begin to fracture. After the yield point, a long plateau of plastic deformation occurs as a result of progressive fracture and collapse of additional trabecular bone and marrow spaces (**Figure 12**). Once the entire marrow

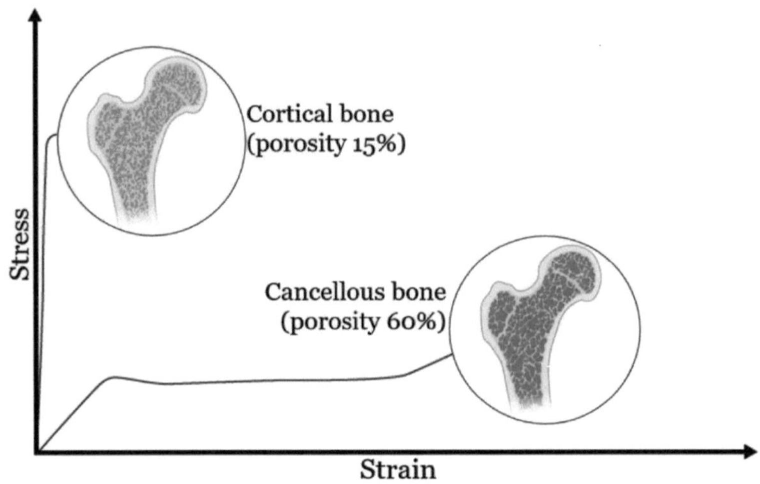

Figure 12.
Different stress/strain curve profiles for cortical and cancellous bone.

space has filled with debris from fractured bone trabeculae, which is referred to as pore closure, there is a marked increase in stiffness before the ultimate failure point of cancellous bone is reached. Under compression loading, cancellous bone exhibits a stress-strain behavior similar to that of soft porous metal. When compression loading is applied, cancellous bone can absorb a large amount of energy (when compared to cortical bone) and can tolerate strain values up to 7% before structural failure·.

On the contrary, cortical bone, due to its low porosity, presents a brittle behavior when subject to compressive loads, similar to glass. Cortical bone is characterized by a decreased plastic deformation phase before failure, absorbs less energy, and tolerates lower strain values (<2%) before fracture as compared with cancellous bone (**Figure 11**). However, cortical bone has greater ultimate strength and increased stiffness and can tolerate more force loads before fracture than cancellous bone.

The clinical implications of the relationship between bone's apparent density and its mechanical behavior are evident when large changes in the strength and modulus of bone can result from small changes in its apparent density. In the clinical setting, the reduction of apparent density is evident on radiographs only when lost by 30–50%, and consequently, the reduction in bone density detected on radiographs is associated with greatly reduced stiffness and strength [1]. Conversely, greatly enhanced fracture zone stiffness and strength may be present even with minor increases in fracture zone density observed in radiographs.

5. Applied fracture biomechanics to common clinical presentations in small animal osteosynthesis

When the concepts of fracture biomechanics are applied to clinical situations, in simple terms, it is possible to define strain as movement and stress as force or a magnitude of load that is applied to the bone and/or the implant.

5.1 Strain

The plate strain is the strain (movement) experienced by a plate when a load of the force vector is applied to it. More specifically, it is the amount of movement that the plate experienced with a certain force (proportional to the original length). Areas of high strain on the plate are areas of high stress. Areas of high plate strain should be avoided because small increases in stress on a plate decrease the fatigue life of an implant. The majority of implant failures after small animal orthopedic surgery are fatigue failures.

It is generally accepted that strain/movement at the fracture gap needs to be within the tolerable levels for tissues. The fracture stability dictates the type of healing that will occur. With strain below 2%, primary bone healing can occur, whereas at 100% strain, the only tissue that can form is granulation tissue. In secondary bone healing, the initial tissue at the fracture site in the inflammatory phase of bone regeneration is granulation tissue. The tissue then progressively stiffens until cartilage can form. Cartilage has a strain tolerance of around 10%. *In vitro* data suggests that callus is stimulated at strains of around 5–10% and bone is stimulated at strains between 1 and 5%. The bone formation starts in lower strain zones at the periphery near the periosteum before spreading inward across the entire fracture gap. .

5.2 Stiffness

The concept of stiffness can be thought of as the magnitude of movement when a force is applied (it is the slope of the stress/strain curve). If the implant is stiff, it does not move when force is applied. One of the determinants of stiffness is working length. If the working length is increased, the stiffness decreases. This means more movement of the plate and higher stress (and higher strain). However, if the working length of the plate decreases, the stress and the strain will be concentrated in a smaller area, which can also predispose to plate failure.

5.3 The strain paradox

Stoffel et al. found that in an *in vitro* situation of a 1 mm simple fracture gap, the strain experienced on the plate in tension bending was lower with a long working length [18]. However, if you have a more flexible plate, the fracture ends touch and suddenly load sharing is produced and therefore less movement and lowered strain, however, only in tension bending. Although with a stiffer plate, the plate does not bend in tension bending. Basically, in the situation of a 1 mm gap, the strain was paradoxically decreased with a longer working length. However, this phenomenon can be explained by the fact that fracture ends touched when the bone is loaded, preventing further movement of the plate in the '*in vitro*' situation. If strain and stress on implants are increased, the fatigue life of a plate decreases. If the fatigue life of a plate from 100,000 cycles is reduced to 10,000 cycles, this could be the difference between the fracture healing before implant failure and catastrophic failure requiring surgical revision. .

5.4 The concept of micromotion

It is widely recognized that micromotion contributes to fracture healing by stimulating the formation of bridging calli. Osteosynthesis methods that are based on relative stability allows micromotion creating a biomechanically optimal construct for secondary bone healing by promoting bone callus formation and has already been associated with early bone healing in several high-risk cases [19]. On the other hand, delayed unions resulting from insufficient mechanical stability, or hypertrophic non-union, may also be associated with large callus formation.

The concept of micromotion is also applied to joint prostheses at the bone-implant interface Excessive micromotion of an implant in bone renders bone ingrowth impossible and reduces osteointegration of prosthesis. The tolerated minimal movement within an interface has been reported to be 28–150 μm, and repetitive higher displacements values allow only the ingrowth of fibrous tissue to avoid osteointegration [20]. Micromotion magnitude is primarily a function of implant stability, although is influenced by the differences in the elastic modulus of bone and implants.

Axial micromotion can be created with circular external skeletal fixators (because the wires allow motion at the fracture site that is axial in direction), with some configurations of interlocking nail and with special plate designs. However, when using a locking plate with a long working length, the micromotion observed at the fracture site is characterized by not only an axial vector; it is also multidirectional.

Besides the influence of the magnitude of micromotion, the characteristics of interfragmentary micromotion are also influential in bone healing. Applying cyclic interfragmentary micromotion for short periods has been shown to influence the repair process significantly [21]. In a study by Goodship et al., it was reported that interfragmentary cyclic micromovement applied for the short term at a high strain rate produced a greater amount of periosteal callus when compared to the same stimulus applied at a low strain rate. It was also shown if a high-strain-rate stimulus is applied later in the regeneration period, this physiological process was significantly inhibited [21]. The beneficial effect of this particular biophysical stimulus early in the healing period may be related to the viscoelastic nature of the differentiating connective tissues in the early endochondral callus. In the early endochondral callus, high rates of movement induce a greater deformation of the fracture fragments because of the stiffening of the callus [19, 21].

An experimental study proved that stimulation of new bone formation by dynamization with micromovement was effective mainly in the early healing phase (4 weeks postoperatively), while dynamization had no significant influence in the late healing phase (8 weeks postoperatively). The beneficial effects of micromotion are hampered by the influence of the gap size in the healing process [22]. From that evidence, with dynamization, the negative effects related to a large gap size overcome the positive effects of dynamization [22]. If a flexible fixation of a simple diaphyseal fracture is performed in clinical practice, the fracture gap should therefore be reduced to as small as possible. But if for some reason a large fracture gap cannot be avoided, dynamization (i.e., enabling micromovement) of the fracture should be performed very carefully and only in the first weeks postoperatively [23]. A large callus formation does not necessarily lead to greater mechanical stability [23, 24]. From that conclusion, was not the size of the radiological evident callus, but the amount of newly formed bone of the peripheral callus that was important for gaining mechanical stability. After the early healing phase, a large amount of new bone is formed, which is mainly responsible for the biomechanical stability of the fracture line.

The amount of callus, more specifically the periosteal callus, is, to some extent, related to the flexural rigidity of the fracture. Research that has found a consistently positive effect of interfragmentary movement on the mechanical stability of regenerating bone has applied only small and controlled interfragmentary movements in the early healing phase [25] or allowed larger movement and loads in a later phase [26].

5.5 Excessive stiffness of implants

The concept that stiffer implants delay bone healing assumes that a callus cannot be formed when strain conditions are too low. In a situation where strain is 0%, potentially this could delay healing. However, it is an unlikely situation in clinical scenarios. When animals use the limb, the amount of force applied always causes a strain value at the fracture site over 0%, and for that reason, there will be an unrealistic complication in small animals. Under optimal stiffness repairs are much more common in veterinary patients, often delayed, and nonunions are a consequence of inadequate addressing of fracture mechanics and/or poor biology versus too stiff implants. Low-strain environments created by stiffer implants facilitate haversian canals and faster bone regeneration.

5.6 The concept of elastic osteosynthesis

This concept is very specific to juvenile dogs and cats. Different breeds of dogs reach skeletal maturity at different ages; it is considered that the physiological process is finished between 5 months (toy breeds) and 18 months (giant breeds) through a very rapid and biphasic growth rate. During the initial growth phase, both structural and material properties of immature bone are considerably different from those of adult bone and are characterized by lower strength and stiffness, as well as lower yield stress and elastic modulus. Additionally, the diaphyseal cortices are thinner but have a more robust periosteum in young animals compared to in adults. As a consequence, immature canine bone is more predisposed to implant failure due to screw pull-out. In addition, due to the rapid initial growth phase and the natural flexion angle of the elbow and knee, postoperative immobilization of these joints in young dogs will inevitably lead to ankylosis secondary to adhesion formation and muscle contracture. In the hind limb, if the functional recovery does not happen early on after osteosynthesis, fracture disease leads to irreversible loss of function due to muscle contracture even after a few days of immobilization. To prevent this debilitating complication, early osteosynthesis is recommended to promote controlled postoperative mobilization, which can lead to implant failure due to the hyperactive nature of non-leash-trained puppies.

The use of overly rigid fixation in juveniles can lead to concentrated forces at the screw-bone interface. In the situation of a standard cortical screw, in this poor-quality soft juvenile bone, this could result in poor screw purchase, screw loosening, and subsequent implant failure, mostly due to screw pull-out. This situation is less common with locking screws, as for a locking screw to fail, it needs to cut through the bone.

Regardless of the osteosynthesis technique chosen and used in juvenile or pediatric dogs, physes must be preserved at all cost. This absolute requirement contraindicates the use of any intramedullary implants (e.g., pins or interlocking nails) especially during the first, rapid growing phase where the physes are more sensitive to traumatic closure. The external fixation is not the technique of first choice for the osteosynthesis of humeral or femoral diaphyseal fractures in young dogs due to mechanical and biological reasons. Namely, the outward position of the external fixator construct, away from the neutral axis of the bone, elevates the bending stresses at the pin/bone interface, promoting a stress riser point. These osteosynthesis technique is also prone to early failure due to implant pull-out, and the use of positive profile transfixation pins does not reduce this complication. From a biological standpoint, the transfixation of the thigh or arm musculature increases the exudation at pin/soft-tissue interface due to excessive movement and generates postoperative pain avoiding free range of motion (ROM) at the knee or elbow. The resulting loss of ROM potentially leads to muscle contracture. Due to the potential complications associated with intramedullary pining and external fixation techniques, plate osteosynthesis remains the treatment of first choice for diaphyseal fractures in juvenile dogs However, if the AO principles of anatomical reduction and rigid internal fixation were used routinely in early growth phase, it can result in catastrophic implant failure via screw pull-out, which leads to the creation of elastic plate osteosynthesis technique (EPO). The technique relies on the increased overall compliance of the bone/plate construct to reduce the risk of focal failure of the screw/bone interface. EPO is used in conjunction with minimally invasive surgical strategies (MIS) favoring restoration of alignment rather than anatomical reconstruction and percutaneous sliding plate techniques to further decrease postoperative morbidity and stimulate early functional recovery. The plates used in EPO were mainly veterinary cuttable plates preferably with locking screws used in a bridging function without anatomical reduction and hematoma disturbance due to their favorable effects on indirect bone healing. Indirect fracture reduction is accomplished by traction on the distal fragment with small fragment forceps and/or using the plate. Large fragments or an oblique fracture should be reduced with the aid of pointed reduction forceps but without attempting a precise reduction. Since anatomical reduction is not attempted; restoration of the bone length is achieved by determining the appropriate plate length from radiographic views of the contralateral intact bone. Since the fracture site is not exposed, it is beneficial to verify proper alignment via intraoperative radiography or fluoroscopy.

The plate is cut to the desired length according to the anticipated position of the screws relative to the growth plates and inserted epiperiosteal through two proximal and distal small incisions. Cortical screws are placed in the two most proximal and the two most distal holes of the plates without tapping to increase bone adherence. In order to decrease pull-out complication, the screws axis should always be oriented in diverging planes in relation to bone longitudinal axis.

The preservation of the strong periosteal sleeve, and the use of an undersized implant such as veterinary cuttable plates (VCP), allow controlled motion at the fracture site, which in turn promotes secondary bone healing via fast callus formation. The flexural or bending deformation of the bone/plate construct is controlled, in part, by the working length of the plate dimension. EPO guidelines recommended that the central plate span without screws should be as long as possible and include no less than 3 consecutive empty screw holes to increase compliance and reduce stress riser effect. This pattern of screw distribution increases the working length of the plate and therefore its compliance. As a result, it decreases the stress riser effect of a single empty screw hole, thus reducing the risk of implant fatigue failure. Furthermore, the enhanced compliance of the bone/plate system lowers the stress on the interface between the bone and screw, thus decreasing the possibility of screw pull-out. Another strategy to decrease screw pull-out complication would be the use cancellous screws instead of cortical screws. The cancellous screw has larger threads and a higher pitch as compared to the cortical screw, which makes its use indicated in metaphyseal bone, osteoporotic bone, or low-porosity bone as found in young patients.

The use of minimally invasive (percutaneous) plate osteosynthesis in conjunction with EPO further reduces postoperative morbidity and promotes early use of the fractured limb and a rapid functional recovery. With this method of osteosynthesis, bone union was achieved as early as two weeks and in all cases at four weeks postsurgically [27]. Surgical complications related to implant failures, such as screw pull-out and plate plastic deformation, were not reported. Radiographically, callus remodeling could be visualized two months postoperatively, and the bony union was completed in four months [27]. Diaphyseal growth occurred without complications, and angular deformation was not observed in either epiphysis.

Although weight-bearing and ROM are recommended immediately after surgery, high-impact activities (jumping and rough play), while difficult to be truly controlled, should be avoided. In contrast, controlled physical activities such as leash walking, trotting, and swimming are beneficial to bone regeneration and should be stimulated.

5.7 Osteosynthesis in toy-breed dogs

In toy dog breeds, complications related to osteosynthesis were more frequently reported than in the general population [28]. Delayed or nonunion and stress protection have been documented in long bone fractures of toy breeds as the most frequent complications, with a special focus on the radius and ulna [29]. Refracture after plate removal is a common complication after stabilization of the radius and ulna fractures. Patient factors such as poor intraosseous vascularity and limited periosseous soft tissue coverage predispose small-breed dogs to healing complications [30].

Biological osteosynthesis techniques decreasing iatrogenic surgical trauma while yielding appropriate construct stability would appear to be advantageous for facilitating the healing of these fractures. External skeletal fixation can be used in toy-breed dogs; however, the radius is a very narrow bone, in addition to its elliptical cross section, which makes the placement of transosseous ESF pins technically challenging [28]. Piras et al. reported the use of circular external skeletal fixators (CESF) in radius and ulna fractures in 16 toy-breed dogs, all of which achieved union despite reporting a 40% minor complication rate, including pin and wire tract discharge [31]. Plate osteosynthesis classically is considered a successful surgical option despite the report of major complications in 18% of cases in one study [32]. Nevertheless, more recently published studies have described a reduction in complications overall or implant-related. Hamilton et al. reported a series of 14 toy-breed dogs treated with a T-plate, all of which healed uneventfully [33]. Regarding function assessment, it was graded as excellent in six dogs, good in four, and fair in two dogs. Vallefuoco et al. only reported 9% of implant-related complications with the use of LCP plates, which could explain the lowering of complications over time [34]. Despite MIPO being recommended in this group of dogs due to poor intraosseous vascularity and limited periosseous soft tissue coverage, recent studies have shown that conventional plate fixation of these fractures is not associated with such a high complication rate when fractures are treated with an appropriately sized bone plate. Pozzi et al. reported a retrospective study that radius and ulna fractures managed with MIPO had similar alignment, reduction, and time to union as fractures managed with ORIF [29]. Arburn et al. also reported a low rate of complications (3%) when ORIF for distal radial fractures was used [35].

In toy breeds, any implant has the potential to lead to stress protection, which can cause osteopenia, especially in radius and ulna fractures. This does not mean that the use of flexible implants is an absolute indication for toy breeds. For the same reasons as above, plates without the appropriate stiffness will fail in the same way as for any dog, especially if the anatomical reduction is required and the fracture line is not uniformly compressed, leaving the transcortices without contact subjecting the plate to bending stress and more prone to fatigue failure.

Excessively rigid plate fixation has historically been considered to be associated with stress protection and subsequent osteopenia, which may in part be responsible for increased refracture risk in these breeds [36, 37]. Osteopenia induced by stress-protection has been reported as a frequent (7.1–20%) complication after plate osteosynthesis of distal radial and ulnar fractures in miniature and toy-breed dogs [28, 32, 38]. A low incidence (1.5%; 1/65) of osteopenia was reported in the study published by Aikawa et al., in part because of the selection of appropriate plate size and type (DCP vs. LCP) with a proper technique [39]. Stress protection-induced osteopenia can only be detected by long-term plate application follow-up [37]; therefore, long-term annual radiographic evaluations are needed to diagnose this complication. On the other hand, a recent study has assigned vascular compromise of the bone cortex as the main cause of osteopenia [40]. Stress protection may not be the cause of osteopenia in distal radial and ulnar fractures, and routine plate removal is not necessary when fractures, provided that plates of appropriate size and type are used and soft-tissue handling atraumatic not overlooked [28, 32]. The diameter of the screws used is another factor to be considered. If they occupy more than 30% of the width of the bone radius (as the maximum size allowed), the bone may have reduced bone strength or have impaired vascular supply, and this can be a reason for osteopenia development [41].

Implant-induced osteoporosis (IIO) or osteopenia can be caused by osteonecrosis of the bone occurring just below the plate that causes cortical bone thinning of about 40%, occurring at 24 weeks after dynamic compression by plate placement [42].

IIO is evolved by biphasic changes and is attributed to inadequate blood supply at 8–12 weeks and reduced mechanical bone stress at 24–36 weeks [37]. IIO is a relatively common complication in small dogs, caused by a process of insufficiently developed bone microvessels, after internal fixation with a conventional plate [43].

LCP plates are reported to preserve blood flow to the periosteum and enable angularly stable fixation, leading to increasingly used in small animal orthopedic surgery [44–46]. In contrast to DCP/LC-DCP in which stability is provided by frictional forces between the plate and bone, locking plates allow the plate to be placed away from the periosteal surface and do not require compression of the periosteum, preserving periosteal blood flow and achieving secondary bone healing due to relative stability [46]. Preserving periosteal blood flow during fracture treatment is an important factor for bone regeneration; as long as the blood flow is preserved, the risk of infection and IIO is reduced. LCP plates due to reportedly small periosteal contact areas reduce the risk of early postoperative osteoporosis and should be the main option for distal radial fractures in toy breeds [47].

Regarding the material used for plating, the comparative studies for the most common alloys used (titanium vs. stainless steel) did not show different results regarding stress shielding [48, 49]. However, titanium alloys produced more flexible plates compatible with the modulus of elasticity of bone. This flexibility is inductive of fracture healing in areas where higher strain values are needed to promote bone regeneration. Additionally, titanium alloy is reported to be more resistant to cyclic load and notch sensitivity when compared to stainless steel and from a theoretical point of view should be the first-choice material for implants used in this type of breed [50].

Plate removal is indicated if osteopenia or IIO is diagnosed due to the predisposition to refractures after implant removal. This procedure should be staged in two to three surgical procedures [51].

5.8 Minimally invasive plate osteosynthesis

Minimally invasive plate osteosynthesis (MIPO) is a surgical approach to fracture treatment using bone plates, following principles that include (1) the use of indirect, closed reduction techniques; (2) epiperiosteal plate insertion through small incisions remote to the unexposed fracture site; and (3) minimal reliance on secondary implants and bone grafts [52].

This surgical approach emphasizes soft tissue preservation over anatomic reconstruction/absolute mechanical stability and is specially indicated for low-strain fractures. In most fractures repaired by MIPO techniques, the bone heals in conditions of relative stability. Relative stability relies on the use of implants that provide flexible fixation, allowing an acceptable degree of strain compatible (<2%) with bone regeneration. Osteosynthesis methods that are commonly used in MIPO are plates or plate-pin combinations applied in bridging function to span a bone defect not anatomically reduced, resulting in a relatively stable environment.

This technique is applicable in the treatment of most diaphyseal, metaphyseal, and periarticular fractures. The use of an intramedullary pin, particularly recommended in comminuted diaphyseal and metaphyseal fractures, is beneficial in facilitating the reduction and restoration of alignment [53]. The minimum recommended diameter for the IM pin is 30% of the medullary canal diameter at the bone isthmus [54].

MIPO is a surgical approach that often ends up with a long working length plate; however, this is because we have chosen to sacrifice the mechanics of our implant, to preserve the biology. This approach can favor the biological factors of bone regeneration, but the increased working length decreases the stiffness of the construct and therefore the fatigue life of the plate. The primary factors affecting the stiffness of the plate are the modulus of the material used, the AMI of the construct, and the working length. The factors influencing gap strain are gap width and the magnitude of motion between the fragments. Fatigue failure are determined by factors such as the yield bending strength of the construct and the cumulative load/number of cycles that are suffered by the plate. The rationale of the MIPO approach is to improve biological factors at the fracture site to speed up healing, preventing plates from prematurely failing due to fatigue failure.

In MIPO, the plate is applied as a bridging function; for that reason, the selection of an implant of appropriate length is a crucial step. With this surgical approach, it is recommended to use longer plates as possible for improving screw-working leverage and to distribute bending forces well along the plate, thereby lowering pull-out forces on screws. If the surgeon chooses the MIPO approach, selecting the adequate plate length in preoperative planning is a crucial step for bridging osteosynthesis. Two parameters are used to determine the plate length: the plate span ratio and the plate screw density. The plate span ratio is the quotient of plate length and segmental length of fractured/comminuted bone. The plate screw density is the quotient of the number of screws inserted and the number of screw holes available. For comminuted fractures, which are commonly treated with MIPO and bridging osteosynthesis, the plate span ratio should be greater than two to three. For simple fractures, this ratio ranges between eight and ten. In comminuted fractures, the plate working length may not be the distance between the screws closest to the fracture, but rather the unsupported area of the plate corresponding to the length of the fracture gap.

Plate screw density or screw-hole-ratio should be smaller than 0.5–0.4 in comminuted fractures and at least two to three screw holes empty over the bone defect [55]. For simple fractures, a value of 0.4–0.3 is recommended. Because this ratio is usually applied to the whole plate, it may not be as applicable for highly comminuted fractures in shorter animal bones. Also, the screw density can be different in different bone segments due to the diversity of lengths, being higher in shorter segments and lower in longer segments. Mechanically, there was a poor advantage of adding more than 4 screws per fragment. Within a fragment, the guidelines advise placing 1 screw close (near) to the fracture and 1 at the very end of the plate (far) and then a minimum of 2 additional screws evenly spaced over the remaining span. Adding more screws offers no mechanical security but does add surgical damage to the bone [2].

The recommended ratio of plate length to bone length [Plate-Bridging Density (PBD)] should be less or equal to 0.91 ± 0.05 [56].

Beyond location, the number of monocortical and bicortical screws in the construct is also influential on its biomechanical properties. Less torsional stiffness is provided with monocortical screws compared to with bicortical screws. When using LCP, a minimum of one screw must be placed bicortically in each major bone fragment due to a significantly increased torsional stability, based on the scientific evidence of a biomechanical study using bone models [18, 57].

Additionally, long plates enable plate insertion incisions to be created far from the fracture site. Surgical planning should include the exact location and sequencing of insertion of the screws to be placed. It is recommended to start inserting the first screw distally to center the plate in the distal segment. To align the bone and stabilize the fracture, the most proximal screw is next inserted into the proximal fracture segment. Additional screws are inserted and used to reduce the bone to the plate. When using a pre-contoured locking plate, it is recommended that a cortical screw be placed in both the distal and the proximal bone segments to frame the bone to the plate, further aligning the bone in the sagittal plane. After stabilizing the fracture with the 2 non-locking screws, locking screws are sequentially placed in the aforementioned order. Preoperative bone plate contouring is advisable to decrease surgical time with the MIPO technique. Preoperative plate contouring can be performed using contralateral bone radiographs or 3D printing models if the contralateral bone is not fractured [58].

An important factor to be considered is the alignment between the bone axis and the plate. Due to poor visualization of the bone surface caused by a limited surgical approach, malalignment between the bone axis and plate leads to an eccentric plate position can occur. At the proximal or distal end of the plate, a monocortical screw will not anchor in the bone [57]. To overcome this problem of insufficient anchorage of a monocortical self-drilling screw, a long bicortical self-tapping screw can be inserted or a standard screw allowing angulation in the plate hole [57]. However, this procedure can also cause iatrogenic fractures [59].

Dynamic compression plate (DCP), limited contact-dynamic compression plate (LC-DCP), or locking compression plate (LCP) systems have been used with success for MIPO procedures. Nowadays, the MIPO technique is almost performed in the majority of cases using a locking plate-screw interface, such as the LCP, due to the angular stability provided by this system, which by definition increases the load-carrying ability of the construct. The angular stability originates from the threaded screw heads being locked into the threaded plate holes, thus forming a fixed-angle construct. Another important advantage of locking plates for use in MIPO is the minimal contouring needed for the application of the plate in contrast to DCP or LC-DCP, which requires optimal contouring to maintain the reduction of the fracture. Locking plates are considered internal fixators and therefore do not displace the fracture segments during screw tightening regardless of the precision of contouring.

The major disadvantages of using monoaxial locking implants are the inability to vary the angle of screw insertion through the hole (unless using a polyaxial locking plate system) and the increased cost of locking implants compared with that of standard plates and screws [58].

On the other hand, non-locking bone plates for MIPO offer other advantages, in radius and ulna fractures where the plate can be used to reduce and align the fracture segments in the sagittal plane. The relatively flat cranial surface of the radius allows precise reduction of the proximal and distal fracture segments as long as the plate has been preoperatively contoured. Many locking plates also allow the insertion of non-locking (cortical) screws into the plate holes (combi holes). If a locking plate is pre-contoured and initially applied to the bone using a cortical screw in the proximal and distal fracture segment, then the locking plate can be used to align the radius in the sagittal plane similar to a non-locking plate. Once sagittal plane alignment is achieved, the remaining screws inserted should be locking screws, to take advantage of the angular stability provided. The cortical screws that were initially inserted may be left in place or replaced by locking screws [58].

LCPs also have the advantage of preserving periosteal vessels. The periosteal blood supply beneath locking plates is not damaged because compression between the plate and the bone does not occur because is not a plate-bone friction base system which improve and hastens bone healing and simultaneously reduced the risk of cortical bone necrosis and infection. Malunion or delayed union are infrequent complications when using this type of implant in MIPO. Regarding infection rates, when MIPO and ORIF are compared, there is a lack of evidence in veterinary studies, but in the human side, evidence showed lower infection rates when MIPO techniques are used in long bone fractures [29, 60–62].

Further advancements with intraoperative imaging such as fluoroscopy have the following aims: maximized biology due to a more limited surgical approach allows placing implants with a longer working length and improve alignment. Alignment of the main bone segments and the articular surfaces without torsional and angular deformities is also one of the main objectives of MIPO. Intraoperatory fracture alignment can be assessed by two methods: intraoperative diagnostic imaging and clinical evaluation. Intraoperative imaging is not always available in clinical practice, and for that reason, precise perioperative planning is a critical point for bone alignment in MIPO.

Fracture reduction under the plate (FRUP) is a technique that was developed by Cabassu et al. to improve bone alignment on MIPO without intraoperative imaging but requires precise preoperative contouring of the plate and extensive preoperative planning [63].

With the FRUP, the first step of surgical planning is to obtain radiographs of the fractured and contralateral bones, under sedation or general anesthesia. Two orthogonal projections of contralateral bone digital radiographs were obtained using a radiopaque marker (of known dimensions) to calibrate images for plate contouring. The choice of the type of fixation is based on fracture location/classification and biological and clinical factors. After calibrating the radiological image, the craniocaudal or mediolateral image of the long bone is used to contour the plate. Ideally, the plate length is selected to span from the proximal to the distal metaphysis of the bone when possible or based on a plate length/fracture length ratio of 3 (MIPO guidelines) [57]. The placement of the plate on the digital radiograph is oriented by anatomical landmarks that could be externally identified intraoperatively such as patella, medial tibial malleolus, femoral greater trochanter, ulna styloid process, lateral epicondyle, and greater tubercle of the humerus. The plate is then anatomically contoured to adapt to the bone surface (e.g., the lateral face of the femur diaphysis and the medial surface of the tibial diaphysis). Fracture line(s) is drawn on the intact bone, which allows planning the number and the position of the screws to be inserted. First, the site to place the screws near the fracture is chosen. According to the MIPO guidelines, at least three empty screw holes should be respected over the fracture site [57].

When managing long oblique or comminuted fractures that have a significant gap, it is recommended to place one screw in each fragment as close to the fracture as possible. In comminuted fractures with a smaller gap, screws are placed with a minimum of three holes' space between them. In the outermost plate holes, one screw is placed in each proximal and distal fragment. Depending on the case, a third screw may be inserted between the inner and outermost screws [63]. The location for each screw is predetermined and identified by its hole number from proximal to distal. The type of screw, whether locking or cortical, is then selected. At least one cortical screw is used on the distal and proximal fragments to allow for fracture reduction, and these screws are placed first in these bone segments [63]. These screws are inserted in the diaphyseal segment of the bone in diaphyseal fractures or close to the fracture site in metaphyseal fractures. Afterward, the surgeon will then subjectively decide whether to place locking or cortical screws based on the screw location and angulation relative to the joint. The order of screw insertion is then selected, starting with the cortical screws used to reduce the fracture. Generally, the first screw inserted in the femur is in the proximal segment, while on the tibia, it is in the distal segment. The reason is that plate location was easiest to determine on these fragments. The cortical screws that were initially inserted may be left in place or removed and replaced by locking screws. Screw length is measured during preoperative planning as well as screw angulation (this is possible using a variable angle locking plate system) to avoid articular penetration. The plate is then sterilized the day before surgery or during patient preparation and draping. Specially designed "L," "Y," or "T" plates have proven to be very useful for MIPO stabilization of distal diaphyseal or metaphyseal fractures of the long bones (especially in radius fractures), which would normally be difficult to stabilize using straight plates [58].

Two skin incisions are made to the level of the bone surface away from the fracture, and an epiperiosteal tunnel is created, and the plate is slid onto the bone surface [64]. Anatomical references are identified, flowing by the alignment of one bone segment with the plate using bone-holding forceps; immediately after this step, the plate is fixed to the bone fragment using the first cortical screw, which is inserted perpendicular to the bone surface [63]. The opposite bone fragment is then distracted using bone-holding forceps to gain length, and alignment in torsional and axial planes, and temporarily stabilized to the plate to maintain alignment and length. Anatomical landmarks on the opposite fragment relative to the plate are checked, and the second cortical screw is inserted to obtain a reduction under the plate. The second cortical screw is then inserted to obtain a reduction under the plate. Alignment is assessed intraoperatively by evaluating the range of motion and alignment of adjacent joints in axial and frontal planes. When an intramedullary pin is used, the fracture is temporarily aligned under the plate and stabilized using bone forceps only to facilitate the intramedullary pin insertion. The pin is then inserted, and correct insertion is assessed by releasing the distal fragment from the plate. If the placement of the pin is evaluated as incorrect, the pin is removed from the distal fragment, and the fragment is manually mobilized to allow placement of the pin in the distal medullary cavity. The plate is then fixed in the same way as without using an intramedullary pin, and other screws are inserted respecting the order of preoperative planning using additional intermediate incisions if necessary. Other screws are then inserted respecting the order of preoperative planning using intermediate incisions when necessary. Intermediate incisions can be used if necessary to verify the alignment of the caudal tibial cortex and medial/lateral cortices of the radius with the plate. When an intramedullary pin is used, the pin is then cut to the appropriate length. At the end of the surgery, postoperative orthogonal radiographs are obtained respecting the preoperative radiographic protocol for the fractured limb.

In conclusion, it is accepted that:

1. More flexible implants increase strain at the fracture site;

2. An increase in working length creates a more flexible implant;

3. If strain and stress on implants are increased, fatigue life decreases;

4. Strain needs to be at tolerable levels for bone formation, and this tends to be very low.

6. Biomechanics of implant biomaterials

Orthopedic implants are commonly used for different types of surgical procedures to gain optimal function and provide stability to bone tissue. When inserting these implants, the characteristics of the material are important for surgical success, and the ideal implant must be biocompatible and nonallergenic from a biological point of view. However, when contoured an implant to the bone surface, its resistance can change significantly. Implants can be temporary or permanent in the body, and metal possesses properties that make it acceptable for bone repair. In orthopedic implants, metals and their alloys were the first materials used in their production, primarily due to their superior strength and biocompatibility. The metals used for implant production include nickel, iron, cobalt, titanium, vanadium, and aluminum. Metal alloys aim to achieve specific properties in the final mixture, such as ductility, strength, elasticity, and corrosion resistance [65]. Ductility is the ability of a material to absorb energy and plastically deform without fracturing. The term ductility is sometimes used to encompass both types of plasticity: tensile (ductility) and compressive (malleability). Current alloys used in orthopedic metal-based implants include stainless steels, cobalt-based alloys, and titanium-based alloys.

6.1 Stainless steel

Stainless steel 18-8 (18% chromium, 8% nickel) is the most common alloy. It has superior corrosion resistance obtained through compositional modifications by using additional metals, especially Cr [66]. The inclusion of Cr allows Cr_2O_3 promotes the formation of a strong and adherent layer that is beneficial for healing. Stainless steel is commonly used in removable orthopedic devices, such as plates, screws, and intramedullary pins, due to its affordability [50, 67]. Currently, the new stainless steel-based alloys contain Co-Cr, Mn, Ni, and a high nitrogen content. Stainless steel alloys have high resistance to corrosion due to their high chromium content (more than 12 wt%), which enables the formation of a strongly adherent, self-healing, and corrosion-resistant coating of Cr_2O_3 oxide. Different types of stainless steel are available for implant production, and the most widely used is austenitic stainless steel. Austenitic stainless steel, which contains austenite-stabilizing elements such as Ni or Mn, is the most commonly used type of stainless steel for implant manufacture. AISI 316L is the most widely used stainless steel in clinical applications, containing 0.03 wt % C, 17–20 wt% Cr, 12–14 wt% Ni, 2–3 wt% Mo, and minor amounts of nitrogen, manganese, phosphorus, silicon, and sulfur [68].

When compared to bone tissue, stainless steel alloys are significantly stiffer and have proven to be durable enough for osteosynthesis [69]. Additionally, stainless steel is relatively inexpensive and biologically well-tolerated, with a smooth surface from electropolishing. It is also ductile enough to allow for contouring of the plate without breaking [69].

6.2 Titanium and titanium-based alloys

Titanium (Ti) and its alloys were initially used in the field of aeronautics but later gained significant interest in the biomedical field due to their remarkable properties. These properties include a moderate elastic modulus of about 110GPa, good corrosion resistance, and low density (around $4700 kgm^{-3}$) [70].

For orthopedic devices, Ti may be used alone or in alloys with other metals, most commonly commercially pure (CP)-Ti and Ti-6Al-4V alloy; this designation refers to its chemical composition of almost 90% titanium, 6% aluminum, 4% vanadium, 0.25% iron (maximum content), and 0.2% oxygen (maximum content). They both provide stable fixation and a low risk of implant loosening [70].

The report of the osseointegration phenomenon for Ti implants by Branemark [71] led to the development of dental and surgical applications of Ti alloys. This property enables titanium and its alloys to tightly integrate with bone, resulting in the improved long-term behavior of the implanted devices, which in turn reduces the risks of loosening and failure.

CP Ti, grade 4 (ASTM F67) and Ti6Al4V (ASTM F136) are the titanium alloys most commonly used for orthopedic implants. For CP Ti-based implants, four grades are currently available varying their oxygen content. CP Ti grade 4 is the type having the highest amount of oxygen (up to 0.4%) and, consequently, the highest tensile and yield strengths [72].

The use of pure titanium has the following advantages: low weight and very good corrosion resistance, especially in saline solution. Ti and its alloys possess outstanding corrosion resistance, which can be attributed to the creation of a robust and adherent TiO_2 oxide layer on their surface. About the surface properties, namely, wear, the performance is poor due to the low shear resistance of Ti and Ti alloys.

The ability to become tightly integrated into the bone greatly improves the long-term mechanical behavior of the implant as well as reduces the risk of loosening and failure of the device [73–75].

CP Ti, with a single-phase alpha microstructure, is currently used for dental implants production, while Ti6Al4V, with a biphasic alpha-beta microstructure, is mostly used in orthopedic implants and prostheses. The Al and V alloying elements stabilize the alpha-beta microstructure and improve the mechanical properties of CP Ti (typically twice the yield and ultimate strength values of CP Ti). Mechanical properties of CP Ti and their alloys can be altered by heat treatment and mechanical working. Although Ti and Ti alloys are characterized by an array of excellent properties (e.g., favorable mechanical characteristics, corrosion resistance, fatigue-corrosion resistance, low density, and relatively low Young modulus), their processing is complex whether it is by machining, forging, or heat treating.

CP Ti and Ti alloys, on the other hand, more closely matches the modulus of elasticity of bone. This flexibility may be more conducive to fracture healing in points where more strain is required for a bone regeneration to develop. Titanium alloy is also more resistant to cyclic loading and notch sensitivity.

6.3 Cobalt-based alloys

Cobalt-based alloys are superior to stainless steel in terms of strength [76]. However, cobalt alloys have better biocompatibility and are more corrosion-resistant. But these alloys are more expensive to produce. Cobalt-chromium-molybdenum alloy variants are specifically used for hip prosthesis implants due to their high abrasion resistance [77, 78].

6.4 Fatigue failure and cyclic loading of implants

Fatigue failure and cyclic loading are two important concepts for guiding the choice of orthopedic implants to avoid construct failure. Clinically, acute deformation or catastrophic failure by a single applied load is a rare event. Several factors can influence the fatigue failure phenomenon such as the magnitude of the applied load (by consequence generates stress within the implant), the geometry of the implant, the material and how it was handled and manufactured, and the local environment of the fracture.

Experimental determination of the fatigue behavior of a material involves creating an S versus N curve, where S represents the applied stress and N represents the number of cycles required for failure (plotted logarithmically). If the applied stress is greater than the yield stress of the implant, the material fails in a few cycles, such as repeatedly bending a paper clip. The number of cycles to cause failure increases as the applied stress is reduced. The stress level at which a material can withstand an infinite number of cycles without failure is called the endurance limit, which is approximately 50% of the ultimate tensile stress for most metals [2]. A similar process can be used to characterize a fatigue behavior of a structure such as a bone plate, applying a load versus number curve. After determining the yield load, a series of progressively decreasing peak loads are established, and the number of cycles required to reach a defined failure point, such as breakage or reduced stiffness, is recorded. The number of cycles required to reach failure increases as the applied load is reduced. An implant's performance may be considered adequate if it survives a clinically relevant number of cycles, which is often set at 10^6 [2].

The response curve for implant construct may be more complex to interpret because geometry, material, and manufacturing factors may all interact. Factors, such as plate screw holes, may cause local stress concentrations that accelerate fatigue failure. The degree of cold working and even the purity of the production process may vary among different manufacturers of similar implants. Macroscopically visible small imperfections and cracks can trigger the implant failure cascade. Surgeons should also be aware that small notches on the surface of a structure can significantly decrease the endurance limit because it is a stress riser and should prompt intraoperative replacement.

In clinical practice, fatigue failure can be avoided by selecting implants of appropriate strength and dimension for the weight and bone size of the animal, minimizing notching, and, through good client compliance to discharge indications, reducing the magnitude and frequency of the applied loads.

Optimizing the rate at which the one regenerates at the fracture gap consolidates also helps avoid fatigue failure because stress in the plate decreases during the regenerative phase of bone due to progressive load sharing.

The local environmental factors related to fractures also can influence fatigue failure of implants. Factors like load sharing between plate and bone, when anatomical reconstruction of fracture is possible (e.g., simple fracture of long bones), highly reduce the cyclic loading magnitude and early failure of the implants [79]. Technical factors related to the correct application of DCP or LC-DCP plates are also crucial to fatigue failure. An illustrative example is the use of DCP plates without pre-bending or overbending of at line fracture. In this case scenario, there will be compression under the plate and distraction on the opposite cortex, causing failure of load sharing and altering strain distribution over the fracture and increasing the magnitude of cyclic loading and fatigue failure of implant more probable [80]. The correct magnitude of pre-bending of the plate is 2 mm prior to a fixation on a convex side of long bones and provides the most compression at the far cortex and consequently the load sharing between bones and plates [80].

LCP has over the years progressively replaced the use of DCP and LC-DCP plates. One of the main advantages of applying this type of implant is the possibility of applying those without adequately contoured and affixed directly to the bone for stable internal fixation of the fracture. For this reason, it has been used in minimally invasive osteosynthesis modalities such as in MIPO and supports biological osteosynthesis by functioning as an internal fixator, rather than as a full (DCP) or limited contact bone plate (LC-DCP) [18, 81]. Additionally, it was reported that LCPs were more resistant to cyclic loading in different force vectors than DCP and LC-DCP [82]. However, to maintain biomechanical advantages, it is advisable that LCP must not be more than 2 mm away from the surface of the bone [81–83].

Bone regeneration in high-strain fractures occurs only if the interfragmentary strain is less than 2%. According to Claes et al., transverse line osteotomies can tolerate up to 2 mm of micromotion without causing harmful damage to bone regeneration [24]. In this type of fracture, anatomical reconstruction is necessary and the strain at the fracture site caused by different force vectors must be neutralized by the implants during the reparative phase of bone regeneration to avoid complications such as delayed or nonunion [84]. With high strain rates, the magnitude and frequency of loading cycles are also greater because the animal will use the limb very early, and the implants will endure a greater number of loading cycles predisposing to fatigue failure. On the other hand, the reparative phase of bone regeneration develops over time, alleviating the magnitude of load cycles due to load sharing.

In low-strain fractures, the interfragmentary movement is not very harmful to the repair process woven bone can tolerate 2–10% of interfragmentary strain [85]. The main objective in this type of fracture is the indirect reduction of bone fragments with bridge plating or external fixation, aiming to re-establish the mechanical axis and bone length and promote secondary bone healing by relative stability [86]. The great advantage of this method is the possibility of a minimally invasive application, and therefore, it is appropriately used in MIPO, where the preservation of the fracture environment is maximized, and bone healing is optimized and even faster than in open reduction and internal fixation (ORIF) [13, 56, 61, 87]. On the other side, from a mechanical standpoint, plates experienced a greater magnitude of strain, increasing the risk of fatigue failure. However, there are surgical options for a sparing effect on

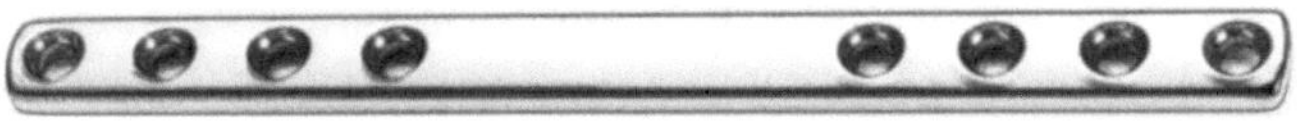

Figure 13.
Image of biological healing plate with a central section without screw holes.

plate site that bridges fracture site and reduce strain. The use of an intramedullary rod (IMR) also helps the restoration of alignment, a substantial challenge in MIPO inherent to the lack of fragment observation and biological healing plates [53, 88]. These former implants are designed to support high strain values for bending and torsional force vectors by possessing a central section without screw holes (**Figure 13**). The screw holes located at the plate's outer section allow the implant to be fixed to the intact proximal and distal fragments, which avoids the need for anatomical reduction of the diaphysis. Additionally, the use of a locking version of this type of plate can improve the performance of the implant by decreasing the pull-out of screws and the need to exactly contour the plate to the bone surface. By applying a MIPO approach, the soft tissue disruption can be minimized, improving biological factors at comminuted fracture sites and hastening bone regeneration.

7. Conclusion

Mastering the concepts of biomechanics in fracture management is an essential tool for the small animal orthopedic surgeon. The application of these concepts in the selection of implants, surgical technique, and fracture healing and their interaction can reduce the rate of postoperative complications. With the rise of minimally invasive osteosynthesis, the knowledge of the most common fracture pattern and the interaction and how the force vectors act on fracture sites determines the choice of implants. On the side of the implant, the knowledge of AMI and the working length of implants determines the yield of the construct and the ability to support the forces before implant failure occurs. Gap strain management is vital for vascular ingrowth and tissue differentiation along the osteogenic pathway. The recognition of the strain pattern at fracture (low strain fracture vs. high strain fracture) is a key element to implant choice and by the influence of the magnitude of the strain at the tissue differentiation (during the osteogenic pathway) also influences fracture healing. Strict adherence to guidelines for implant placement is another pathway to fulfilling evidence-based biomechanics in orthopedic surgery. Finally, an important part of the postoperative assessment of constructs is for surgeons to use their understanding of these mechanical parameters to predict the weakest point and have this guide patient management decision.

Acknowledgments and funding

Ana Catarina Sousa (SFRH/BD/146689/2019) and Bruna Lopes (2021. 05265.BD), acknowledge Fundação para a Ciência e Tecnologia (FCT) for financial support. Rui Damásio Alvites acknowledges the Animal Science Studies Centre (CECA), Agroenvironment, Technologies and Sciences Institute (ICETA), Porto University (UP), and FCT for the funding and availability of all technical, structural, and human resources necessary for the development of this work. The author Patrícia Sousa acknowledges the University of Porto (UP) and Centro de Estudos de Ciência Animal (CECA), Instituto de Ciências, Tecnologias e Agroambiente (ICETA) for the funding (H2Cure: POCI-01-0247-FEDER-047032) and availability of all resources needed for this work. The work was supported through the project UIDB/00211/2020 funded by FCT/MCTES, national funds. This research was funded by Projects PEst-OE/AGR/UI0211/2011 from FCT and COM-PETE 2020, from ANI–Projetos ID&T Empresas em Copromoção, by the project "Print-on-Organs–Engineering bioinks and processes for direct printing on organs" with the reference PO-CI-01-0247-FEDER-033877, and by the project "Bone2Move- Development of "in vivo" experimental techniques and modeling methodologies for the evaluation of 4D scaffolds for a bone defect in sheep model: an integrative research approach" with the reference POCI-01-0145-FEDER-031146.

Conflicts of interest

The authors declare no conflicts of interest regarding the publication of this article.

Author details

Nuno Alexandre[1,2*], Gonçalo Simões[1,2], António Martinho Lopes[4,5], Tarcísio Guerra Guimarães[3], Bruna Lopes[6,7,8], Patrícia Sousa[6,7,8], Ana Catarina Sousa[6,7,8], Rui Damásio Alvites[6,7,8,9] and Ana Colette Maurício[6,7,8]

1 Departamento de Medicina Veterinária, Escola de Ciências e Tecnologia (ECT), Universidade de Évora (UE), Évora, Portugal

2 MED—Mediterranean Institute for Agriculture, Environment and Development & CHANGE—Global Change and Sustainability Institute, Departamento de Medicina Veterinária, Escola de Ciências e Tecnologia (ECT), Universidade de Évora (UE), Évora, Portugal

3 MED—Mediterranean Institute for Agriculture, Environment and Development & CHANGE—Global Change and Sustainability Institute, Institute for Advanced Studies and Research (IIFA), Universidade de Évora (UE), Évora, Portugal

4 Departamento de Clínica, Faculdade de Medicina Veterinária (FMV), Universidade de Lisboa (UL), Avenida da Universidade Técnica, Lisboa, Portugal

5 Centro de Investigação Interdisciplinar em Sanidade Animal (CIISA), Faculdade de Medicina Veterinária (FMV), Universidade de Lisboa (UL), Avenida da Universidade Técnica, Lisboa, Portugal

6 Departamento de Clínicas Veterinárias, Instituto de Ciências Biomédicas de Abel Salazar (ICBAS), Universidade do Porto (UP), Porto, Portugal

7 Centro de Estudos de Ciência Animal (CECA), Instituto de Ciências, Tecnologias e Agroambiente da Universidade do Porto (ICETA), Porto, Portugal

8 Associate Laboratory for Animal and Veterinary Science (AL4AnimalS), Lisboa, Portugal

9 Cooperativa de Ensino Superior Politécnico e Universitário (CESPU), Gandra PRD, Portugal

*Address all correspondence to: nmla@uevora.pt

References

[1] Radasch RM. Biomechanics of bone and fractures. The Veterinary Clinics of North America. Small Animal Practice. 1999;**29**(5): 1045-1082, v-vi

[2] Roe S. Biomechanics of fracture fixation. The Veterinary Clinics of North America. Small Animal Practice. 2020; **50**(1):1-15

[3] Minto BW, Lucena DVF. Bases biomecânicas das fraturas. In: Minto BW, Dias LGGG, editors. Tratado de ortopedia de cães e gatos. 1ª ed. São Paulo: Editora MedVet; 2022. pp. 134-148

[4] Yamashita J, Li X, Furman BR, Rawls HR, Wang X, Agrawal CM. Collagen and bone viscoelasticity: A dynamic mechanical analysis. Journal of Biomedical Materials Research. 2002; **63**(1):31-36

[5] Wu K, Li B, Guo JJ. Fatigue crack growth and fracture of internal fixation materials in in vivo environments—A review. Materials. 2021;**14**(1):176

[6] Stock JT. Wolff's law (bone functional adaptation). In: Trevathan W, Cartmill M, Dufour D, Larsen C, Rourke DO, Rosenberg K, Strier K, editors. The International Encyclopedia of Biological Anthropology. New Jersey, USA: Wiley-Blackwell; 2018. pp. 1-2

[7] Amber, Carter KP, Cheng K, Greenbaum A, Ligler FS, Moatt A. Enhancement of bone regeneration through the converse piezoelectric effect, a novel approach for applying mechanical stimulation. Bioelectricity. 2021;**3**(4):255-271

[8] Ahn AC, Grodzinsky AJ. Relevance of collagen piezoelectricity to "Wolff's law": A critical review. Medical Engineering & Physics. 2009;**31**(7):733-741

[9] Palmer RH, James SP. Fracture biomechanics of the appendicular skeleton. In: Bojrab MJ, Monnet E, editors. Mechanisms of Disease in Small Animal Surgery. 3rd ed. Teton Newmedia; 2016. pp. 150-170

[10] Edgerton BC, An KN, Morrey BF. Torsional strength reduction due to cortical defects in bone. Journal of Orthopaedic Research. 1990;**8**(6): 851-855

[11] Ferguson J. Fracture of the humerus after cancellous hone graft harvesting in a dog. Journal of Small Animal Practice. 1996;**37**(5):232-234

[12] Kallianpur N, Singh K, Gopinathan A, Sarangom SB, John C, Sowbharenya C, et al. Investigation on relation between factors affecting occurrence and outcome of repair of long bone fractures in 216 dogs. International Journal of Livestock Research. 2018;**8**(2): 225-234

[13] Mair JJ, Belkoff SM, Boudrieau RJ. An ex vivo mechanical evaluation of single versus double semitubular plate fixation of a transverse distal-third scapular osteotomy in the dog. Veterinary Surgery. 2003;**32**(6):580-584

[14] Hulse D, Hyman D. Fracture biology and biomechanics. In: Slatter D, editor. Textbook of Small Animal Surgery. Philadelphia: Saunders; 1993

[15] Frank JD, Ryan M, Kalscheur VL, Ruaux-Mason CP, Hozak RR, Muir P. Aging and accumulation of microdamage in canine bone. Bone. 2002;**30**(1):201-206

[16] Lin X, Patil S, Gao Y-G, Qian A. The bone extracellular matrix in bone formation and regeneration. Frontiers in Pharmacology. 2020;**11**(757):1-15

[17] Wall A, Board T. The compressive behavior of bone as a two-phase porous structure. In: Banaszkiewicz PA, Kader DF, editors. Classic Papers in Orthopaedics. London: Springer London; 2014. pp. 457-460

[18] Stoffel K, Dieter U, Stachowiak G, Gachter A, Kuster MS. Biomechanical testing of the LCP–How can stability in locked internal fixators be controlled? Injury. 2003;**34**(Suppl. 2):B11-B19

[19] Han Z, Wu J, Deng G, Bi C, Wang J, Wang Q. Axial micromotion locking plate construct can promote faster and stronger bone healing in an ovine osteotomy model. Frontiers in Bioengineering and Biotechnology. 2020;**8**:593448

[20] Cross MJ, Spycher J. 9—Cementless fixation techniques in joint replacement. In: Revell PA, editor. Joint Replacement Technology. Sawston, UK: Woodhead Publishing; 2008. pp. 190-211

[21] Goodship AE, Cunningham JL, Kenwright J. Strain rate and timing of stimulation in mechanical modulation of fracture healing. Clinical Orthopaedics and Related Research. 1998;**355**(Suppl): S105-S115

[22] Yamaji T, Ando K, Wolf S, Augat P, Claes L. The effect of micromovement on callus formation. Journal of Orthopaedic Science: Official Journal of the Japanese Orthopaedic Association. 2001;**6**:571-575

[23] Claes L, Augat P, Suger G, Wilke HJ. Influence of size and stability of the osteotomy gap on the success of fracture healing. Journal of Orthopaedic Research. 1997;**15**(4):577-584

[24] Claes LE, Heigele CA, Neidlinger-Wilke C, Kaspar D, Seidl W, Margevicius KJ, et al. Effects of mechanical factors on the fracture healing process. Clinical Orthopaedics and Related Research. 1998;**355**(Suppl): S132-S147

[25] Kershaw CJ, Cunningham JL, Kenwright J. Tibial external fixation, weight bearing, and fracture movement. Clinical Orthopaedics and Related Research. 1993;**293**:28-36

[26] Noordeen MH, Lavy CB, Shergill NS, Tuite JD, Jackson AM. Cyclical micromovement and fracture healing. Journal of Bone and Joint Surgery. British Volume (London). 1995;**77**(4): 645-648

[27] Cabassu J-P. Elastic plate osteosynthesis of femoral shaft fractures in young dogs. Veterinary and Comparative Orthopaedics and Traumatology. 2001;**14**(01):40-45

[28] Ramírez J, Macías C. Conventional bone plate fixation of distal radius and ulna fractures in toy breed dogs. Australian Veterinary Journal. 2016; **94**(3):76-80

[29] Pozzi A, Hudson CC, Gauthier CM, Lewis DD. Retrospective comparison of minimally invasive plate osteosynthesis and open reduction and internal fixation of radius-ulna fractures in dogs. Veterinary Surgery. 2013;**42**(1):19-27

[30] Welch JA, Boudrieau RJ, DeJardin LM, Spodnick GJ. The intraosseous blood supply of the canine radius: Implications for healing of distal fractures in small dogs. Veterinary Surgery. 1997;**26**(1):57-61

[31] Piras L, Cappellari F, Peirone B, Ferretti A. Treatment of fractures of the distal radius and ulna in toy breed dogs with circular external skeletal fixation: A retrospective study. Veterinary and Comparative Orthopaedics and Traumatology. 2011;**24**(3):228-235

[32] Larsen LJ, Roush JK, McLaughlin RM. Bone plate fixation of distal radius and ulna fractures in small- and miniature-breed dogs. Journal of the American Animal Hospital Association. 1999;**35**(3):243-250

[33] Hamilton MH, Langley Hobbs SJ. Use of the AO veterinary mini 'T'-plate for stabilisation of distal radius and ulna fractures in toy breed dogs. Veterinary and Comparative Orthopaedics and Traumatology. 2005; **18**(1):18-25

[34] Vallefuoco R, Le Pommellet H, Savin A, Decambron A, Manassero M, Viateau V, et al. Complications of appendicular fracture repair in cats and small dogs using locking compression plates. Veterinary and Comparative Orthopaedics and Traumatology. 2016; **29**(1):46-52

[35] De Arburn PR, Benamou J, Gatineau M, Clerfond P, Planté J. Open reduction and cranial bone plate fixation of fractures involving the distal aspect of the radius and ulna in miniature- and toy-breed dogs: 102 cases (2008-2015). Journal of the American Veterinary Medical Association. 2017;**250**(12): 1419-1426

[36] Gilbert JA. Stress protection osteopenia due to rigid plating. Clinical Biomechanics (Bristol, Avon). 1988;**3**(3): 179-186

[37] Field JR. Bone plate fixation: Its relationship with implant induced osteoporosis. Veterinary and Comparative Orthopaedics and Traumatology. 1997;**10**(02):88-94

[38] Kang BJ, Ryu HH, Park S, Kim Y, Kweon OK, Hayashi K. Clinical evaluation of a mini locking plate system for fracture repair of the radius and ulna in miniature breed dogs. Veterinary and Comparative Orthopaedics and Traumatology. 2016;**29**(6):522-527

[39] Aikawa T, Miyazaki Y, Shimatsu T, Iizuka K, Nishimura M. Clinical outcomes and complications after open reduction and internal fixation utilizing conventional plates in 65 distal radial and ulnar fractures of miniature- and toy-breed dogs. Veterinary and Comparative Orthopaedics and Traumatology. 2018; **31**(03):214-217

[40] Gauthier CM, Conrad BP, Lewis DD, Pozzi A. In vitro comparison of stiffness of plate fixation of radii from large- and small-breed dogs. American Journal of Veterinary Research. 2011;**72**(8):1112-1117

[41] Johnson AL, Houlton JE, Vannini R. AO Principles of Fracture Management in the Dog and Cat. Stuttgart, Germany: Georg Thieme Verlag; 2005

[42] Slätis P, Karaharju E, Holmström T, Ahonen J, Paavolainen P. Structural changes in intact tubular bone after application of rigid plates with and without compression. The Journal of Bone and Joint Surgery. American Volume. 1978;**60**(4):516-522

[43] Kessler SB, Deiler S, Schiffl-Deiler M, Uhthoff HK, Schweiberer L. Refractures: A consequence of impaired local bone viability. Archives of Orthopaedic and Trauma Surgery. 1992; **111**(2):96-101

[44] Gibert S, Ragetly GR, Boudrieau RJ. Locking compression plate stabilization

of 20 distal radial and ulnar fractures in toy and miniature breed dogs. Veterinary and Comparative Orthopaedics and Traumatology. 2015; **28**(06):441-447

[45] Haaland PJ, Sjöström L, Devor M, Haug A. Appendicular fracture repair in dogs using the locking compression plate system: 47 cases. Veterinary and Comparative Orthopaedics and Traumatology. 2009;**22**(04):309-315

[46] Muroi N, Ochi H, Shimada M, Asou Y, Hara Y. Effects of long-term plate fixation with different fixation modes on the radial cortical bone in dogs. PLoS One. 2021;**16**(2):e0247410

[47] Greiwe RM, Archdeacon MT. Locking plate technology–current concepts. The Journal of Knee Surgery. 2007;**20**(01):50-55

[48] Joshi G, Naveen B. Comparative study of stainless steel and titanium limited contact-dynamic compression plate application in the fractures of radius and ulna. Medical Journal of Dr DY Patil Vidyapeeth. 2019;**12**(3):256-261

[49] Zhang S, Patel D, Brady M, Gambill S, Theivendran K, Deshmukh S, et al. Experimental testing of fracture fixation plates: A review. Proceedings of the Institution of Mechanical Engineers, Part H: Journal of Engineering in Medicine. 2022;**236**(9):1253-1272

[50] Barber CC, Burnham M, Ojameruaye O, McKee MD. A systematic review of the use of titanium versus stainless steel implants for fracture fixation. OTA International. 2021; **4**(3):e138

[51] Muroi N, Shimada M, Murakami S, Akagi H, Kanno N, Suzuki S, et al. A retrospective study of postoperative development of implant-induced osteoporosis in radial-ulnar fractures in toy breed dogs treated with plate fixation. Veterinary and Comparative Orthopaedics and Traumatology. 2021; **34**(6):375-385

[52] Gerber C, Mast JW, Ganz R. Biological internal fixation of fractures. Archives of Orthopaedic and Trauma Surgery. 1990;**109**(6):295-303

[53] Hulse D, Ferry K, Fawcett A, Gentry D, Hyman W, Geller S, et al. Effect of intramedullary pin size on reducing bone plate strain. Veterinary and Comparative Orthopaedics and Traumatology. 2000;**13**(04):185-190

[54] Pearson T, Glyde MR, Day RE, Hosgood GL. The effect of intramedullary pin size and plate working length on plate strain in locking compression plate-rod constructs under axial load. Veterinary and Comparative Orthopaedics and Traumatology. 2016; **29**(06):451-458

[55] Strauss EJ, Schwarzkopf R, Kummer F, Egol KA. The current status of locked plating: The good, the bad, and the ugly. Journal of Orthopaedic Trauma. 2008;**22**(7):479-486

[56] Schmokel HG, Stein S, Radke H, Hurter K, Schawalder P. Treatment of tibial fractures with plates using minimally invasive percutaneous osteosynthesis in dogs and cats. The Journal of Small Animal Practice. 2007; **48**(3):157-160

[57] Gautier E, Sommer C. Guidelines for the clinical application of the LCP. Injury. 2003;**34**(Suppl 2):B63-B76

[58] Hudson CC, Lewis DD, Pozzi A. Minimally invasive plate osteosynthesis: Radius and ulna. The Veterinary Clinics of North America. Small Animal Practice. 2020;**50**(1):135-153

[59] Namm JD, Morris RP, Speck FL 3rd, Lindsey RW. The impact of eccentric diaphyseal plate and screw placement on the risk of peri-implant fracture. The Journal of Bone and Joint Surgery. American Volume. 2018;**100**(20): 1765-1770

[60] Boero Baroncelli A, Peirone B, Winter MD, Reese DJ, Pozzi A. Retrospective comparison between minimally invasive plate osteosynthesis and open plating for tibial fractures in dogs. Veterinary and Comparative Orthopaedics and Traumatology. 2012; **25**(5):410-417

[61] Pozzi A, Risselada M, Winter MD. Assessment of fracture healing after minimally invasive plate osteosynthesis or open reduction and internal fixation of coexisting radius and ulna fractures in dogs via ultrasonography and radiography. Journal of the American Veterinary Medical Association. 2012; **241**(6):744-753

[62] Marazzi C, Wittauer M, Hirschmann MT, Testa EA. Minimally invasive plate osteosynthesis (MIPO) versus open reduction and internal fixation (ORIF) in the treatment of distal fibula Danis-Weber types B and C fractures. Journal of Orthopaedic Surgery and Research. 2020;**15**(1):1-9

[63] Cabassu J. Minimally invasive plate osteosynthesis using fracture reduction under the plate without intraoperative fluoroscopy to stabilize diaphyseal fractures of the tibia and femur in dogs and cats. Veterinary and Comparative Orthopaedics and Traumatology. 2019; **32**(06):475-482

[64] Schmierer PA, Pozzi A. Guidelines for surgical approaches for minimally invasive plate osteosynthesis in cats. Veterinary and Comparative Orthopaedics and Traumatology. 2017; **30**(4):272-278

[65] Shekhawat D, Singh A, Bhardwaj A, Patnaik A, editors. A short review on polymer, metal and ceramic based implant materials. In: IOP Conference Series: Materials Science and Engineering. IOP Publishing; 2021

[66] Wall EJ, Jain V, Vora V, Mehlman CT, Crawford AH. Complications of titanium and stainless steel elastic nail fixation of pediatric femoral fractures. JBJS. 2008;**90**(6): 1305-1313

[67] Mahyudin F, Widhiyanto L, Hermawan H. Biomaterials in orthopaedics. In: Biomaterials and Medical Devices: A Perspective from an Emerging Country. 2016. pp. 161-181

[68] Navarro M, Michiardi A, Castano O, Planell JA. Biomaterials in orthopaedics. Journal of the Royal Society Interface. 2008;**5**(27):1137-1158

[69] Marin E, Boschetto F, Pezzotti G. Biomaterials and biocompatibility: An historical overview. Journal of Biomedical Materials Research Part A. 2020;**108**(8):1617-1633

[70] Filip N, Radu I, Veliceasa B, Filip C, Pertea M, Clim A, et al. Biomaterials in orthopedic devices: Current issues and future perspectives. Coatings. 2022; **12**(10):1544

[71] Branemark PI, Breine U. Formation of bone marrow in isolated segment of rib periosteum in rabbit and dog. Blut. 1964;**10**:236-252

[72] Hoffman AS, Schoen FJ, Lemons JE. Biomaterials Science: An Introduction to Materials in Medicine. Cambridge, USA: Academic Press; 2004

[73] Baltatu MS, Tugui CA, Perju MC, Benchea M, Spataru MC, Sandu AV, et al. Biocompatible titanium alloys used in medical applications. Revista de Chimie. 2019;**70**(4):1302-1306

[74] İzmir M, Ercan B. Anodization of titanium alloys for orthopedic applications. Frontiers of Chemical Science and Engineering. 2019;**13**: 28-45

[75] Rony L, Lancigu R, Hubert L. Intraosseous metal implants in orthopedics: A review. Morphologie. 2018;**102**(339):231-242

[76] Mehta H, Kaur G, Chaudhary GR, Prabhakar N, Kaul S, Singhal NK. Evaluation of corrosion resistant, antimicrobial and cytocompatible behaviour of cobalt based metallosurfactants self-assembled monolayers on 316L stainless steel surface. Surface and Coatings Technology. 2022;**444**:128657

[77] Thanigaivel S, Priya A, Balakrishnan D, Dutta K, Rajendran S, Soto-Moscoso M. Insight on recent development in metallic biomaterials: Strategies involving synthesis, types and surface modification for advanced therapeutic and biomedical applications. Biochemical Engineering Journal. 2022; **108522**:1-11

[78] Ramakrishna S, Mayer J, Wintermantel E, Leong KW. Biomedical applications of polymer-composite materials: A review. Composites Science and Technology. 2001;**61**(9):1189-1224

[79] Chao P, Conrad BP, Lewis DD, Horodyski M, Pozzi A. Effect of plate working length on plate stiffness and cyclic fatigue life in a cadaveric femoral fracture gap model stabilized with a 12-hole 2.4 mm locking compression plate. BMC Veterinary Research. 2013;**9**(1):125

[80] Ristow J, Mead M, Cordeiro M, Ostrander J, Atkinson T, Atkinson P. Pre-bending a dynamic compression plate significantly alters strain distribution near the fracture plane in the mid-shaft femur. Proceedings of the Institution of Mechanical Engineers, Part H: Journal of Engineering in Medicine. 2020;**234**(5):478-485

[81] Ahmad M, Nanda R, Bajwa AS, Candal-Couto J, Green S, Hui AC. Biomechanical testing of the locking compression plate: When does the distance between bone and implant significantly reduce construct stability? Injury. 2007;**38**(3):358-364

[82] Aguila AZ, Manos JM, Orlansky AS, Todhunter RJ, Trotter EJ, Van der Meulen MC. In vitro biomechanical comparison of limited contact dynamic compression plate and locking compression plate. Veterinary and Comparative Orthopaedics and Traumatology. 2005;**18**(4):220-226

[83] Perren SM. Physical and biological aspects of fracture healing with special reference to internal fixation. Clinical Orthopaedics and Related Research. 1979;**138**:175-196

[84] Mavčič B, Antolič V. Optimal mechanical environment of the healing bone fracture/osteotomy. International Orthopaedics. 2012;**36**(4):689-695

[85] Glatt V, Evans CH, Tetsworth K. Reverse dynamisation: A modern perspective on Stephan Perren's strain theory. European Cells & Materials. 2021;**41**:668-679

[86] Alcântara BM, Minto BW, Franco GG, Lucena DVF, Dias LGGG. Bridge plating for simple tibial fractures treated by minimally invasive plate osteosynthesis. Arquivo Brasileiro de

Medicina Veterinaria e Zootecnia. 2021; **73**(03):589-597

[87] Perren SM. The technology of minimally invasive percutaneous osteosynthesis (MIPO). Injury. 2002;**33** (Suppl 1):VI, VII

[88] Guiot LP, Déjardin LM. Prospective evaluation of minimally invasive plate osteosynthesis in 36 nonarticular tibial fractures in dogs and cats. Veterinary Surgery. 2011;**40**(2):171-182

Chapter 3

Advances in Clinical Application of Bone Mineral Density and Bone Turnover Markers

Junyan Li, Niuniu Yuan, Huizhen Wang and Wang Qingzhong

Abstract

Bone mineral density is the main basis for the diagnosis of osteoporosis. The measurement methods of bone mineral density include dual X-ray absorptiometry (DXA), quantitative computer tomography (QCT), quantitative ultrasound (QUS), magnetic resonance imaging (MRI) and so on. Currently, bone mineral density measured by dual-energy X-ray absorptiometry (DXA) is the gold standard for the diagnosis of osteoporosis. Bone turnover markers (BTMs) are biochemical products that reflect the activity of bone cells and the metabolic level of bone matrix, and they reflect the dynamic changes of bone tissue in the whole body earlier than bone mineral-density, procollagen type 1 N-terminal propeptide (PINP) and carboxy-terminal cross-linked telopeptide of type 1 collagen (CTX) is sensitive BTMs, widely used in clinical practice, and can predict the occurrence of fractures. Some new markers such as Periostin, AGEs/RAGE, Gelsolin, and Annexin A2 provide new clues for exploring the mechanism of osteoporosis. The combination of the two can better carry out the diagnosis and differential diagnosis of multiple metabolic bone diseases, evaluate the therapeutic response of anti-osteoporotic medicines, and predict fracture risk.

Keywords: osteoporosis, bone mineral density, bone turnover markers, Periostin, dual X-ray absorptiometry (DXA)

1. Introduction

Osteoporosis (OP) is a systemic bone disease characterized by low bone mass and damage to the microstructure of bone tissue, causing increased bone fragility and susceptibility to fractures [1]. With the aggravation of the global population aging, the prevalence of osteoporosis and the associated fractures is increasing year by year [2]. The medical care and nursing produced by that require a lot of human, material and financial investment, rising serious consequences for families and society such as the huge economic burden and social pressure [3]. Therefore, osteoporosis has become an important public health problem around the world, and early diagnosis is of critical significance for the prevention and treatment of osteoporosis [4, 5]. The diagnosis of osteoporosis is frequently based on bone mineral density, while bone turnover markers were used for differential diagnosis, observation of curative effect and treatment follow-up.

2. Advanced imaging assessments of bone mineral density

Bone mineral density refers to the amount of bone contained in a unit volume (volume density) or a unit area (area density). There are many methods of bone mineral density measurement, and different methods have different roles in the diagnosis of osteoporosis, monitoring of curative effect and assessment of fracture risk. Plain film absorptiometry (RA) and single-energy X-ray absorptiometry (single x-ray absorptiometry. SXA) two detection methods have been rarely used in clinical practice. X-ray plain film can evaluate changes in bone mineral density, but its sensitivity and accuracy are not high. It's difficult to make a positive diagnosis when bone mineral loss is less than 20%.Only when the bone mass is reduced by more than 30%, or even more than 50%, there are abnormal manifestations [6], thus it is generally not used as a tool for routine evaluation of bone mineral density. Presently, the commonly used bone mineral density measurement methods in clinical and scientific research include dual energy X-ray absorptiometry (DXA), quantitative computed tomography (QCT), quantitative ultrasound (QUS) and MRI, etc.

2.1 DXA

Dual X-ray Absorptiometry (DXA) measures two-dimensional bone mineral density (areal BMD), namely the measured bone mineral content (bone mineral content, BMC) divided by the measured bone projection area. DXA bone mineral density measurement is the most commonly used method for bone mineral density measurement in clinical and scientific research. The main measurement site is the axial bone, including: lumbar spine and proximal femur. Lumbar BMD can sensitively reflect the changes of bone metabolism and therapeutic effect, while femoral BMD is one of the most reasonable indicators for predicting femoral fractures. Anterior and posterior lumbar spine measurements are generally selected for lumbar BMD examination, and the region of interest includes the vertebral body and its posterior appendages. The regions of interest for proximal femur measurement were the BMD of the femoral neck, greater trochanter, total hip and Wards triangle, and the regions of interest for the diagnosis of osteoporosis were the femoral neck and total hip. If the measurement of the lumbar spine and proximal femur is limited, especially when secondary osteoporosis (e.g., hyperparathyroidism) is considered, the non-dominant distal forearm third (33%) can be selected. The distal forearm measurement can obtain the bone mass parameters of the radius, ulna, and radius plus ulna at the super-distal end, the distal mid-segment, the distal 1/3, and the total distal part, totaling 12 different regions [7].

BMD measured by DXA is currently a common diagnostic index for osteoporosis. For postmenopausal women and men aged 50 and over, the BMD value according to the diagnostic criteria recommended by WHO is lower than the peak bone value of healthy adults of the same sex and race. The patients with T value less than 1 are considered as healthy; the T value ranging from 1 and 2.5 as osteopenia (or low bone mass); the T value equal to or more than 2.5 are diagnosed as osteoporosis, the patients with severe osteoporosis usually have one or more fragility fractures simultaneously (**Table 1**). Bone mineral density is usually expressed by T-Score, T-value = (measured value - peak bone mineral density in normal young people of the same race and sex)/standard deviation of peak bone mineral density in normal young people of the same race and sex. For children, premenopausal women and men under the age of 50, it is recommended to use the Z value of the same race to judge the level of bone mineral density, z-value = (bone mineral density measurement value - the

Disease state	T value
normal	T Value ≥ -1.0SD
osteopenia	-2.5SD < T value <−1.0SD
osteoporosis	T value ≤ -2.5SD
Severe osteoporosis	T value ≤ -2.5SD combined with a fragility fracture

Table 1.
The diagnostic criteria for osteoporosis based on DXA, BMD, and T values.

mean bone mineral density of the same race and the same sex and the same age) /the same race and the same age. The standard deviation of bone mineral density among sex peers, z-values below −2.0 were considered as "low cohort expected range" or low bone mass.

The lumbar spine BMD examination generally chooses the anterior and posterior lumbar spine measurement, and the area of interest includes the vertebral body and its posterior appendage structures, so the measurement results are affected by the degenerative changes of the lumbar spine (such as bone hyperplasia and sclerosis of the vertebral body and vertebral facet joints, etc.), abdominal Aortic calcification, intervertebral disc calcification, schmorl node, etc. Literature studies suggested that the choice of lateral lumbar spine BMD measurement can avoid the interference of the above factors [8]. At the same time, about 60% of the vertebral body is cancellous bone, which is also a site prone to osteoporotic compression fractures, while the spinous process, transverse process and pedicle of the posterior 1/3 of the spine are rich in cortical bone, which can be difficult for osteoporotic compression fractures and not play an important role in fractures. Lateral measurement of the lumbar spine can exclude the posterior 1/3 of the spine and detect early vertebral bone loss. In addition, with aging, the bone loss of cortical bone and cancellous bone is different. During a person's lifetime, BMD of the anterior vertebral body decreases by about 50%, while the posterior decreases by about 25%. Therefore, the lateral BMD measurement of the vertebral body can better reflect the actual changes of the spongy bone and the bone mass of the vertebral body itself. The lateral lumbar spine bone mass measurement is paired (accompanied) with the anterior and posterior lumbar spine, that is, combined with the lateral scan measurement on the basis of the anterior and posterior scan measurements, so that the estimated volumetric bone mineral density (vBMD) of the lumbar two-dimensional scan can be obtained at the same time. Also known as width-adjusted BMD (WA-BMD), the bone mass parameters of each vertebral body and the entire vertebral body can be obtained. It also avoids some interference factors and improves the ability of early detection of bone loss, thereby improving the diagnosis of bone loss and susceptibility to loose tissue. The lateral thoracolumbar vertebral images collected by the DXA measuring instrument can also be used for vertebral morphological assessment and vertebral fracture assessment (VFA), but the repeatability of DXA lateral lumbar spine measurement is not as good as the anteroposterior one.

Although BMD measured by DXA is currently recognized as the gold standard for the diagnosis of osteoporosis, there are still some limitations. DXA has the characteristics of high specificity and low sensitivity for the prediction of fracture, and depends on the choice of diagnostic point. A large number of studies have shown that BMD only partially reflects bone strength and cannot effectively evaluate the effect of anti-osteoporosis treatment. It only partially reflects changes in bone structure during aging, metabolic disorders or treatment. More scholars began to pay attention to how to expand the DXA measurement function: 1) Trabecular bone score (TBS) is a measurement index for evaluating bone microstructure by analyzing image pixels of lumbar spine DXA [9]. 2) Hip structure analysis (HSA) is to evaluate the bone strength by computer analysis of the geometric data obtained from the DXA scan image of the proximal femur [10]. 3) Finite element analysis (FEA) is a two-dimensional model for evaluating femoral strength parameters, which can be used as a hip fracture risk assessment [11]. 4) Body composition measurement, which can be used for the evaluation of body composition, and can provide information on BMC, bone density, lean mass and fat content in different regions of the body, but the whole body bone density cannot be used for diagnosis of osteoporosis [12]. 5) Bone density assessment around the prosthesis, DXA can evaluate the stability of the prosthesis by measuring the bone density around the prosthesis [13].

DXA is a currently widely used technology with low radiation dose, and highly recognized as bone mineral density measurement method, while there are still many deficiencies. The regional BMD measured by DXA is a comprehensive measurement of cancellous bone and cortical bone, and the measurement results cannot reflect the early changes in BMD. At the same time, due to the principle of DXA plane projection imaging technology, the area BMD measured by DXA is affected by weight, scoliosis, bone hyperplasia, vertebral fractures and vascular calcification and then reduce the accuracy of BMD measurement. Testing in pregnant women is not yet recommended. As development of osteoporosis percentage increasing and the research of DXA new function in the elderly, further improvement of DXA fan beam scanning technology and application of multidetector, the scope of the application of the low radiation dose DXA is expanding in the assessment of human body bone mineral density measurement. But in addition to the DXA bone mineral density measurement, body composition analysis and evaluation are relatively mature, other functions (such as HAS, TBS, FEA detection, peripheral bone mineral density measurement, etc.) are mostly limited to the preliminary clinical application or the research phase of the trial.

2.2 Quantitative ultrasound (QUS)

QUS is a non-ionizing technology for BMD detection using acoustic waves, which uses different parameters to reflect the situation of bone mass indirectly [14]. Since Longton et al. (2008) first used QUS to measure bone tissue in 1984.The theory, methods, and instruments for measuring BMD with QUS have been greatly developed [15]. There are four types of US transmissions: trabecular transverse transmission, cortical transverse transmission, cortical axial transmission, and pulse-echo measuring devices [16–18]. Among them, Trabecular Transverse Transmission is mainly used to measure cancellous bone and the detection site is calcaneus. Cortical Axial Transmission is used for cortical bone detection and detection site is Radius [19]. Other measurements sites of QUS devices are finger phalanges, tibia, less common femur, posterior processes of the spine and ulna. Through QUS, two parameters are mainly obtained: Broadband Ultrasound Attenuation (BUA) and Speed of Sound (SOS). In theory, the two QUS principal variables are both related not only to BMD but also to trabeculae orientation, the proportion of trabecular and cortical bone, the composition of organic and inorganic components, bone elasticity damage and fatigue [20]. But currently the extent of its impact on BUA and SOS is unknown. The correlation between QUS parameters and DXA-BMD is good. It can distinguish patients with osteoporosis from normal people, but the false negative rate is high. At present, there is no uniform standard for the diagnosis of osteoporosis by ultrasonography. It is not appropriate to apply the WHO diagnostic standard of T ≤ −2.5SD, and its sensitivity and specificity are not ideal. Trimpou_ et al. (2019) pointed out that QUS quantitative ultrasound measurement is mainly attenuation of ultrasonic signals caused by reflection and absorption of sound waves by structures in the region of interest (including soft tissue, bone tissue, and bone marrow tissue).

QUS measurements are not only correlated with bone mineral density, but also provide information about bone stress, structure and more. It is currently mainly used for screening of osteoporotic risk populations and risk assessment of osteoporotic fractures in clinical routine. Several studies of original, review or meta-analyses settings demonstrated that heel QUS parameters are strong predictors of osteoporotic fractures [21–25]. The ultimate clinical use of heel QUS parameters to assess the fracture risk will have to be based and further validated in currently widely used approaches such as FRAX.

QUS has some advantages like simplicity, no radiation damage, high repeatability, low price, and easy handling, etc., Also, QUS can be used in children and pregnant women for primary osteoporosis screening and fragility fracture prediction. Especially in medical facilities where DXA or QCT is deficient, bone density measured by quantitative ultrasound is not true for bone mineral content. It cannot yet be used for the diagnosis of osteoporosis and the judgment of drug efficacy. At present, there is no unified QUS screening judgment standard and it can be referred to the information provided by QUS equipment manufacturers. In addition, horizontal comparison of equipment from different manufacturers cannot be carried out. If the results were suspected for osteoporosis, further DXA measurements should be performed. In conclusion, although QUS currently has recognized limitations in clinical practice, it has also been widely used, especially in the field of pediatrics, township health centers, and physical examination and screening structures. Besides, substantial progress has been made [26]. The parameters of the device for evaluating bone quality are a good supplement to DXA, and it needs to be further standardized before it can be promoted clinically [14, 27].

2.3 Quantitative computer tomography (QCT)

QCT is a method of bone mineral density measurement based on CT scan data after QCT phantom calibration and professional software analysis [14]. QCT uses CT three-dimensional volume data for analysis, and measures the true volumetric bone mineral density (vBMD), which can more sensitively reflect changes in bone BMD. Compared with DXA, QCT measurement is not affected by spinal hyperplasia and regression. The influence of factors such as changes and vascular calcification can avoid the false negative results of planar projection bone mineral density measurement technology caused by the above factors [28]. At the same time, the raw data of QCT can also be used for complex image processing to analyze and study bone changes and structural features [29].QCT includes central QCT, peripheral QCT and high-resolution peripheral bone quantitative CT (HR-pQCT), and micro-CT.

2.3.1 Central QCT (central computed quantitative tomography, cQCT)

cQCT is a pattern that uses multiple two-dimensional slices, the central delineation area of the pattern is the lumbar spine (especially the L1–3 vertebral bodies), the proximal femur, and central QCT also provides a measure of muscle mass [30]. Compared with DXA, central QCT is a measure of mean volumetric BMD (mg/cm3), which improves the sensitivity and accuracy of BMD measurement and can assess the biological properties of interosseous BMD, bone geometry, and bone strength [31]. However, its disadvantage is that it increases the load of ionizing radiation, and due to the fact that most scanners are single-energy devices, which will lead to the potential problem of bone marrow fat changes. Studies have shown that the sensitivity of lumbar spine QCT to determine BMD is better than that of lumbar spine and hip DXA measurement, and it can more accurately reflect the changes in bone metabolism [32]. Clinical needs to choose to do spine or hip. Hip CT scans can be used for QCT, and the measured BMD results are equivalent to DXA areal BMD [33, 34]. According to the diagnostic criteria of the International Society for Clinical Bone Densitometry (ISCD) and the American College of Radiology (ACR), studies have found that QCT is more sensitive than DXA to detect osteoporosis [35]. This diagnostic criterion applies to postmenopausal women and older men. Lumbar vertebra QCT diagnostic criteria for osteoporosis: taking the average value of cancellous BMD of 2 lumbar vertebrae (usually the first and second lumbar vertebrae), and using the absolute value of lumbar spine QCT BMD for diagnosis. The evidence of BMD larger than 120 mg/cm^3 usually is classified as normal, the absolute value of BMD in the range of 80–120 mg/cm^3 as the group of low bone mass, the absolute value of BMD less than 80 mg/cm^3 being considered as osteoporosis [36].

2.3.2 Peripheral quantitative computed tomography (pQCT)

The measurement sites of pQCT are the distal radius and tibia, and the measurement results at this site mainly reflect the cortical bone mineral density. With a low radiation burden compared to central QCT, this modality not only provides valuable data on volumetric BMD, interseptal BMD, bone geometry, and bone strength, but also provides data including cross-sectional area and muscle density, which can be used to assess the risk of hip fractures in postmenopausal women. Because there is no diagnostic standard at present, it cannot be used for the diagnosis of osteoporosis and the judgment of clinical drug efficacy.

2.3.3 High-resolution peripheral computed tomography (HR-pQCT)

HR-pQCT is newly developed QCT scanning modality, which can reconstruct multiple 2D slices (most commonly the radius or tibia) into a 3D virtual bone biopsy and provide enhanced spatial resolution beyond that provided by cQCT, pQCT or MRI [37]. The effective radiation dose of standard HR-pQCT in the distal radius or tibia is 3–5 μsv, which is considered to be a low radiation dose examination compared with other common medical imaging techniques [38]. HR-pQCT assessments have been performed in large epidemiological cohort studies such as the MrOs, OFELY, CaMos and Framingham Osteoporosis Study, which notably can be used for in vivo bone microstructural imaging at peripheral bone sites to understand the pathophysiology underlying bone fragility and improve fracture prediction. The pathophysiological is the basis of fragility and improve the prediction of fractures [39, 40]. And HR-pQCT is based on semi-automatic profiling and segmentation of tissue, which provides data from density, morphology, microstructure, and biomechanical (including stiffness and elastic modulus) measurements through finite element analysis. The clinical application and research of HR-pQCT in many other metabolic diseases exceeds osteoporosis, such as drug effects, rare bone diseases, hand joint imaging and fracture healing. It is used in rheumatoid arthritis to assess joint space width and bone erosion, in knee osteoarthritis and in some studies of fracture healing of the distal radius [41, 42]. The unique advantage of HR-pQCT is the high spatial resolution in vivo, which enables the quantification of trabecular and cortical bone microstructure. HR-pQCT has high research value in bone quality, especially microstructure [43]. However, HR-pQCT is expensive and the imaging technology needs to be further standardized. Although recent recommendations for standardization in scanning, analysis, quality control, and result reporting have been given, the prospect of HR-pQCT in clinical practice still needs to be further studied [44].

There are some advantages and disadvantages for the QCT diagnostic measurements. The main advantages included the followings: ①The measurement of true volumetric bone mineral density is not affected by bone size and shape; ② Selective measurement of cancellous bone mineral density, more sensitive to reflect the changes of early bone mass; ③ The 3D geometric measurement parameters can be used to measure the bone mineral density of multiple sites and analyze the bone composition of cross sectional image; ④ It can be used in preoperative evaluation of orthopedics to guide the selection of clinical surgical methods and surgical sites. The disadvantage of DXA is not as common as DXA in clinical application because of its large size, expensive examination, larger dose of radiation received by patients and smaller application range than DXA.

In conclusion, QCT has been widely used in the clinical and health management of osteoporosis in recent years due to its advantages in imaging technology. Although QCT is more accurate in measuring volumetric bone density, it can measure cortical bone density separately bone and cancellous bone density, while the radiation is larger and there is a partial volume effect. In the vast majority of clinical cases, patients are undergoing CT scans for medical reasons, and the QCT bone mineral density analysis system is used to simultaneously scan the patients to obtain bone mineral density values, without additional radiation doses for patients. QCT can also measure intra-abdominal fat and liver fat content, and QCT combined with low-dose chest CT has a promising application in health management [45, 46].

2.4 Magnetic resonance imaging (MRI)

MRI uses strong magnetic fields and electromagnetic pulse sequences to obtain three-dimensional images. It has the advantages of sensitive signal display and rich post-processing. It can perform quantitative bone density examination, and can also perform bone microstructure imaging to understand the internal situation of bone structure, especially in judging osteoporotic fractures, it is superior to X-ray and CT examination, and there is no X-ray radiation. In recent years, various MR imaging techniques have gradually highlighted their advantages in the field of osteoporosis research, mainly including the followings [47, 48]. 1) Transverse relaxation time ($T2^*$) measurement is a quantitative MRI that indirectly reflects the morphological structure of bone tissue through the $T2^*$ value of the bone marrow. Due to the difference in magnetic susceptibility between trabecular bone and bone marrow tissue, the magnetic field at the junction between the two is not uniform, and the morphological and structural changes of bone trabecular bone will affect the relaxation characteristics of the surrounding bone marrow. In the gradient echo sequence, the bone marrow $T2^*$ value changes. And it has a certain order of magnitude relationship with the number of trabecular bone. Studies have shown that MRI $T2^*$ values are moderately inversely correlated with quantitative computed tomography to assess bone mineral density in postmenopausal women with osteoporosis, and have certain potential in assessing the severity of lumbar osteoporosis [49]. A large number of studies have confirmed that T2* is closely related to osteoporosis, but its sensitivity, specificity, random type, parameters and many other reasons are different [50]. Currently, there is no standard for the diagnosis of osteoporosis with T2*. 2) High resolution MR (HRMR) HRMR scanning has been widely used in recent years. The imaging is based on the signal difference between bone marrow and trabecular tissue. In the background of high signal in the bone marrow, trabecular bone appears as a black network structure. Studies have shown that the bone structure parameters of HRMR have a good correlation with the morphological structure parameters of tissue slices at the same site. The HRMR scanning matrix can reach the order of microns, which can better observe the trabecular bone microstructure and diagnose osteoporotic fractures [51–53]. The effect of HRMR in the detection of osteoporosis is positive, while MR examination time is relatively long, the price is high, and the evaluation is relatively complicated. There is still a lot of work to be done, such as sensitivity, specificity, accuracy, and standardized data processing. At present, it is not widely used in clinical practice, but is believed that with the deepening of research and the improvement of MR software and hardware. MR imaging will definitely play an important role in the diagnosis of osteoporosis. 3) Magnetic resonance spectroscopy (MRS) MRS can evaluate the organic matter, inorganic matter and bone matrix density of bone. Currently, there are phosphorus spectroscopy (13P-MRS) and hydrogen proton spectroscopy (^{1}H-MRS). Among them, phosphorus spectroscopy is to use the echo signal of ^{13}P in bone to determine the content of bone inorganic components [54]. ^{1}H-MRS uses chemical shift to detect bone marrow water and adipose tissue, analyze its biochemical composition and metabolic changes, and indirectly assess bone quality from the molecular level [55]. Due to high technical requirements and many influencing factors, MRS has not been widely used in clinical evaluation of osteoporosis. 4) Others diffusion-weighted imaging (DWI)reflects the early changes in bone marrow composition and can quantitatively assess bone marrow changes. The apparent diffusion coefficient (ADC) and signal-to-noise ratio (SIR) can better reflect the bone mineral density of vertebral bodies in patients with lumbar spine diseases, and can quantitatively evaluate them, which is important for the diagnosis of lumbar spine osteoporosis [56, 57]. Diffusion tensor imaging (DTI) characterizes the diffusion direction of water molecules, which is helpful in assessing fracture risk in patients with osteoporosis [58]. Perfusion-weighted imaging (PWI) uses paramagnetic contrast agents to induce transient changes in the local magnetic field of perivascular tissue, which can reflect the perfusion and hemodynamic changes in tissue microcirculation, and help to detect early abnormal blood supply in diseased tissue [58].

MRI has a good auxiliary role in the diagnosis and differential diagnosis of osteoporosis by taking advantage of its multi-sequence imaging. Tomography can be used to understand the internal situation of the bone structure, Bone quality can be evaluated quantitatively, noninvasively and without radiation. It can reflect the physiological and pathological changes of bone histologically, and better understand the physiological characteristics of bone, so as to make its diagnosis more early and accurate. Because the image analysis process and parameter thresholds of HR MR and quantitative magnetic resonance (QMR) examinations have not been unified, functional imaging such as diffusion-weighted imaging (DWI) and diffusion tensor imaging (DTI) are very important for osteoporosis. The significance of the diagnosis is inconclusive, and the MRI examination is expensive and time-consuming. Therefore, QMR in the diagnosis of osteoporosis still needs further research. With the further maturity of MR imaging technology, the further improvement of coils and the application of higher field strength MR machines, it will be possible to optimize the MR imaging of trabecular bone structure and make the technology of MRI evaluation of trabecular bone structure more mature.

2.5 Comparison of various imaging examination techniques

In conclusion, the above imaging techniques have their own emphasis: DXA has been widely used to evaluate BMD because of its economy, simplicity and low radiation. What's more, WHO has also recommended it as the "gold standard" for diagnosing OP. Although QCT is more accurate in measuring volume BMD, it can measure cortical bone and cancellous bone BMD respectively, but the radiation is larger. QUS is simple and radiation-free, which is mainly used as a screening tool for osteoporosis. MRS is radiation-free and can indirectly assess bone quality at the molecular level. On the premise of bone mineral density measurement, MR combined with QCT or QUS for the detection of osteoporosis, the combined application of multiple methods enhances our scientific understanding of bone microstructure, bone geometric properties and other biomechanics, and provides a basis for further exploration of osteoporosis. The pathophysiological process of the disease, sensitive clinical diagnosis, monitoring of disease changes and curative effects provide technical support (**Table 2**).

Project	Detection of parts	parameters	Clinical application
plain x-ray film	Vertebrae, wrist, metacarpal, calcaneus and tubular bone	—	The sensitivity and accuracy of bone mineral density evaluation are poor, but it can be used to locate the fracture
DXA	Spine, hips, distal forearm, whole body	Areal bone mineral density	It is currently recognized as the gold standard for the diagnosis of osteoporosis and can be used for body composition analysis
QUS	calcaneal, Radius, finger phalanges, tibia	BUA,SOS	It is mainly used for osteoporosis screening
QCT		Volumetric bone mineral density	It can distinguish cortical bone from cancellous bone and diagnose osteoporosis. It is more sensitive to fracture, especially fine fracture
cQCT	Lumbar vertebrae and proximal femur	Mainly cancellous bone mineral density	It can be used to diagnose osteoporosis
pQCT	Radius and tibia	Mainly cortical bone mineral density	To assess the risk of fracture
HR-pQCT	Radius and tibia	Mainly cortical bone mineral density	To quantify the bone microstructure and improve the prediction of fracture
MRI	Refer to QCT site	The related parameters of bone microstructure were evaluated indirectly	It can perform bone microstructure imaging, which is mainly used for differentiating microfracture, new fracture and bone tumor

Table 2.
Comparison of imaging techniques for various bone mineral density examinations.

3. Research progress in bone turnover markers

The diagnosis of osteoporosis also requires etiological diagnosis to further distinguish primary or secondary [59]. Bone turnover markers provide an important reference for clinical differential diagnosis and treatment follow-up. Bone tissue continuously undergoes bone modeling and bone remodeling to maintain bone growth and structural integrity. The microenvironment is characterized by continuous absorption of old bone to form new bone. This self-renewal process is called bone turnover (bone turnover). Bone turnover biomarkers (BTMs) are biochemical markers released in blood or urine during bone remodeling, which can reflect the dynamic changes of whole body bone tissue earlier than bone density. Including biochemical markers of bone formation and bone resorption, the former reflects the activity of osteoblasts and the state of bone formation, and the latter represents the activity of osteoclasts and the level of bone resorption. The determination of these markers is helpful for identifying primary and secondary osteoporosis, judging the type of bone turnover, predicting the rate of bone loss, assessing fracture risk, understanding disease progression, selecting interventions, monitoring drug efficacy and compliance, etc.

The common clinical biochemical markers of bone metabolism are shown in the table below (**Table 3**). Among the above markers, the International Osteoporosis Foundation (IOF) and International Federation of Clinical Chemistry (IFCC) have recommended serum P1NP and CTX-1 as bone formation and resorption reference markers, which have the characteristics of good sensitivity, high specificity, good repeatability, and economical application. In recent years, with the deepening of research and the development of biotechnology, the research on bone metabolism markers has made great progress. New markers such as Periostin, advanced glycation end products/receptor for advanced glycation end products (AGEs/RAGE), gelsolin, annexin A2 etc. gradually emerged, which has potential advantages in reflecting the dynamic changes of the whole body bone tissue.

Bone formation markers	Bone resorption markers
alkaline phosphatase, ALP	tartrate-resistant acid phosphatase, TRACP
osteocalcin, OC	serum C-terminal telopeptide of type 1 collagen, S-CTX
bone alkaline phosphatase, BALP	urinary pyridinoline, Pyr
procollagen type 1 C-peptide, P1CP	urinary deoxypyridinoline, D-Pyr
procollagen type 1 N-peptide, P1NP	urinary C-terminal telopeptide of type 1 collagen, U-CTX
	urinary N-terminal telopeptide of type 1 collagen, U-NTX

Table 3.
Common clinical biochemical markers of bone turnover.

3.1 Procollagen type 1 N-peptide (P1NP)

Osteoblasts synthesize a large amount of type I procollagen, and its carboxyl and amino termini extend to both ends respectively to form the precursor of type I collagen. The propeptides extending toward the carboxyl end are C-propeptides. During bone formation, type I procollagen is secreted to the outside of the cell, and is cleaved 1: 1:1 into procollagen type I N-peptide (PINP), type I collagen and procollagen type I C-peptide (PICP), mature type I collagen mainly constitutes the main component of osteoid, while PINP and PICP enter into the blood and urine as metabolites, so the detection of PINP and PICP can reflect the level of bone formation [60, 61]. Both P1NP and P1CP are metabolized in the liver. Because the half-life of serum PICP fluctuates greatly, the research evidence that P1NP reflects bone formation is more abundant than that of P1CP. Clinically, it is recommended to use PINP as an indicator of bone formation to reflect the synthesis rate of type I collagen and bone turnover [62].

Most studies suggest that elevated PINP can predict fractures. In postmenopausal osteoporotic patients, the P1NP of fracture patients is significantly higher than that of non-fracture groups, and PINP can be used as an important indicator to predict postmenopausal osteoporotic fractures [63]. A meta-analysis of postmenopausal women and men over 50 years of age showed that the hazard ratio (HR) of osteoporotic fractures was 1.18 for every one standard deviation increase in serum PINP [64]. Further studies have shown that high PINP is primarily associated with spine and hip fracture risk, predicting fractures with greater accuracy in the short term (5 years) than in the long term (10 years or more) [65]. After adjusting for BMI, smoking, frequency of falls, previous fracture history, vitamin D intake and other confounding factors, the Crandall study included 800 postmenopausal women with an average follow-up time of 7.13 years was found that serum PINP levels were not correlated with the risk of incidence of hip fractures [66]. Another Meta-analysis, with a total of 11,572 participants, showed that serum PINP levels were not significantly associated with fractures before confounding factors were adjusted. After adjusting for confounding factors (including age, BMI, previous fracture history and BMD, etc.) Raising one standard deviation level, the HR for fracture was 1.28. Whether PINP has a predictive effect on fracture occurrence is still inconsistent due to different statistical methods and different confounding factors in PINP research. At the same time, studies have found that PINP has a good predictive effect on the occurrence of fractures in non-diabetic patients, but has no predictive effect on the occurrence of fractures in patients with type 2 diabetes, suggesting that PINP will have different effects on fracture prediction under different health conditions [66]. Therefore, the correlation between PINP and osteoporotic fractures still needs to be further confirmed by large sample and prospective studies.

3.2 C-terminal telopeptide of type 1 collagen (CTX)

In the process of bone resorption, the mature type I collagen is cleaved and the C-terminal peptide and N-terminal peptide are removed. The common C-terminal peptides are α-CTX and β-CTX, which are isomers, and their production rate is equal to the degradation of type I collagen. CTX and NTX are released into the blood with the degradation of type 1 collagen molecules and can be excreted in the urine. Therefore, the concentrations of CTX and NTX in the blood and urine can specifically reflect the activity of osteoclasts and the level of bone resorption [67]. Since β-CTX has been studied more as a marker of bone resorption, it is clinically used as a sensitive and specific marker of bone resorption [68]. At the same time, CTX-I showed a circadian rhythm, and its concentration peaks usually appeared at night and early morning, and reached the lowest point in the afternoon [69]. And for the measurement of CTX-I, food intake has a greater impact on the results, so it is necessary to measure CTX-I in a fasting state [70].

A number of studies on women have suggested that elevated β-CTX is associated with fracture risk. Vilaca (2017) found that for each standard deviation increase in serum β-CTX, the risk of vertebral fractures increased by 1.4–2.2 times, and the risk of non-vertebral fractures increased by 1.8–2.5 times, and the results were basically unchanged after adjusting for BMD, indicating that CTX has an independent predictive effect on fracture risk [66]. Fracture risk is better predicted if CTX is combined with BMD. The Swedish EPIDOS study showed that the 10-year fracture risk of postmenopausal women from high to low was as follows: ① Elevated serum β-CTX + history of fragility fracture; ② Elevated β-CTX + T value of BMD lower than - 2.5; ③ BMD Women with a T value below −2.5 + a history of fragility fracture; ④ elevated β-CTX or a history of fragility fracture; ⑤ BMD T value below −2.5 [66]. CTX may have a good application prospect in predicting the occurrence of osteoporotic fractures. However, it is still difficult to popularize and apply in clinical practice, and the results are still uncertain due to the high heterogeneity among different studies. Therefore, further large-sample, homogeneous prospective studies are still needed for detailed clarification in the future.

3.3 Periostin

Periostin is a newly discovered macromolecular glycoprotein. As a unique extracellular matrix protein, it is mainly expressed in the periosteum, also known as bone-specific factor 2 which is obtained from the osteoblast cell line MC3T3-E1 cDNA library by Takeshita et al. (1993). A bone adhesion have a molecular weight of 90-kDa [71]. Periostin mainly triggers signaling pathways such as NF-KB/STAT3, P13K/Akt and focal adhesion kinase (FAK) by binding to cell surface integrin receptors αvβ3 and αvβ5, and regulates the expression of downstream genes. It plays an important role in adhesion, tissue repairing and maintaining the integrity of connective tissue structure and function [72].

Basic research suggests that Periostin can regulate bone formation, promote bone development/remodeling, and increase bone strength. It is a key regulator of bone microstructure and plays a very important role in bone metabolism [73, 74]. Regarding the clinical study of periostin, Li et al. (2021) showed through cross-sectional observation in postmenopausal women that periostin has no significant correlation with the overall BMD [75], but is positively correlated with cortical bone density, negatively correlated with cortical bone porosity. Periostin is primarily responsible for periosteal metabolism, so it is more closely related to long bones covered by periosteum and can better reflect cortical bone loss [76]. Further studies suggested that periostin was not associated with baseline BMD and was significantly elevated in women with fractures [77–79]. Kim emphasized that it was primarily a risk factor for nonvertebral fractures [80]. Rousseau proposed that periostin is an independent risk factor for fractures in postmenopausal women, and microarray analysis suggested that periostin mRNA was up-regulated twice in the process of osteoporosis and fracture repairing [77].

In conclusion, periostin, as a new-generation biochemical marker of bone metabolism, is an independent risk factor for fractures among postmenopausal women. Combined with bone mineral density testing, it can better evaluate and predict the risk of osteoporosis and fracture in patients, and provide a theoretical basis for early intervention.

3.4 Advanced glycation end products/receptors for advanced glycation end products

Advanced glycation end products (AGEs) are a variety of compounds produced by non-enzymatic reactions between reducing sugars (such as glucose) and certain metabolites (such as Methylglyoxal) and protein amino groups [81]. The receptor for advanced glycation end products (RAGE) can be expressed in osteoblasts, osteoclasts and osteocytes [82, 83]. In recent years, studies have found that AGEs/RAGE can cause essential changes in osteoblasts, osteoclasts, and osteocytes, resulting into imbalances in bone remodeling, decreased bone strength, and increased incidence of fractures, which may provide unique diagnosis and treatment ideas and molecular targets for the diagnosis and treatment of osteoporosis [84]. Clinical studies have found that the correlation between sRAGE and bone mineral density is controversial. Studies have found that serum sRAGE levels are significantly higher in postmenopausal women with osteoporosis and low bone mass than those with normal bone density, and sRAGE levels are associated with increased fracture risk [85]. RAGE was positively correlated with bone formation markers P1NP and osteocalcin in elderly men, and this correlation was more significant in men with diabetes [86]. However, there was no significant difference in RAGE levels between postmenopausal women with type 2 diabetes and the control group. There was no significant correlation between serum RAGE levels and bone mineral density, fracture prevalence, and bone turnover markers in the type 2 diabetes group [87]. The research and development of bone tissue engineering, it has been found that AGEs/RAGE can affect the structure and biomechanical properties of bone through various mechanisms. It may have a potential diagnostic role in monitoring osteoporosis, especially the progression of diabetic bone metabolism, but its clinical application is less studied, and its value in predicting fracture risk needs to be further studied [88].

3.5 Gelsolin (GSN)

Gelsolin is a calcium-dependent actin-binding protein that cleaves, caps, and nucleates actin to regulate cytoskeleton structure, cell movement and metabolic processes, and also participates in regulation of cell signal transduction and apoptosis [89]. As an actin-binding protein involved in the assembly and movement of osteoclast cell feet. GSN deficiency can hinder the assembly of osteoclast cell feet and increase bone mass and bone strength. Furthermore, GSN can hinder the assembly of osteoclasts to the bone matrix through integrins activation, thereby ultimately activating osteoclasts and promoting bone resorption [90]. Therefore, in different clinical studies, the relationship between GSN and BMD is not consistent. A Mexican study found that serum GSN levels were reduced in postmenopausal women with low bone mass and osteoporosis but the difference between groups was not statistically significant [90]. Peripheral blood mononuclear cells (PBMs), as precursors of osteoclasts, produce cytokines important for osteoclast development and play an important role in bone metabolism. A cytoplasmic proteomic analysis of PBMs from Caucasian men with very high and very low BMD found that GSN expression was significantly increased in patients with very low BMD [91]. The same study of more than 6000 subjects with very high and very low bone density samples found that there was no significant difference in plasma GSN between men with very high and very low bone density, but GSN levels in postmenopausal women were higher than the extremely low BMD group, and it was negatively correlated with hip BMD [92]. Deng et al. (2014) also found that GSN protein and mRNA levels in the PBM of subjects with low BMD were down-regulated, and SNP rs767770 was only significantly correlated with hip BMD in female Caucasians, suggesting that GSN is an important gene affecting hip BMD in female Caucasians [93]. A study on the correlation between GSN and BMD in Chinese postmenopausal women found that the GSN level in postmenopausal women was significantly higher than that in premenopausal women, and compared with the normal BMD group, the plasma GSN level in the low bone mass or osteoporosis group was significantly higher. There is a negative correlation between plasma GSN and hip BMD in postmenopausal women, and GSN is an independent influencing factor of femoral neck and lumbar spine BMD [94]. In conclusion, the current research shows that plasma GSN may be used as a biochemical marker of bone resorption for the diagnosis of osteoporosis, but more in-depth and extensive research is still needed.

3.6 Annexin A2

Annexin A2 (ANXA2) is a calcium-dependent phospholipid-binding protein expressed on the surface of peripheral blood monocytes, which can stimulate monocytes migration across endothelial cells and osteoclasts formation. However, ANXA2's role in bone remodeling is not limited to osteoclast formation, but can also promote the proliferation and differentiation of bone precursor cells, thereby affecting bone formation [95]. Increased expression of ANXA2 was found in postmenopausal Caucasian women patients with low bone mass and osteoporosis. A recent study found that compared with patients without fractures, the expression of ANXA2 protein in the PBMs of patients with osteoporotic fractures was significantly increased and plasma ANXA2 were inversely related to hip BMD in older population, which are significantly higher in the patients with very low BMD than those in very high BMD [96]. These studies suggest that ANXA2 may be a potential biochemical marker for osteoporosis, but there are few clinical studies on ANXA2.Thus, further longitudinal studies are needed to determine whether plasma ANXA2 levels can predict osteoporosis.

4. Conclusion

In conclusion, bone mineral density has been always regarded as gold standard for the diagnosis of osteoporosis in the world. Biochemical markers of bone metabolism can reflect bone remodeling earlier and have the advantages of non-invasiveness and timeliness. The combination of them can be used for better diagnosis and differential diagnosis of metabolic diseases, drug development, and clinical monitoring of osteoporosis treatment efficacy. In recent years, with the research progress of imaging technology and biological science, it has provided technical support for further detection of bone microstructure, bone geometric properties, and bone strength, and provided a theoretical basis for exploring bone physiology and the pathogenesis of metabolic bone diseases. For the newly developed imaging technology and newly discovered bone metabolism markers, the clinical research evidence is limited, and its safety, specificity, sensitivity, stability and other characteristics in clinical application still need more in-depth and extensive research.

Author details

Junyan Li[1], Niuniu Yuan[1], Huizhen Wang[2] and Wang Qingzhong[2]*

1 Department of Endocrinology and Metabolism, Changzhi Medical College Affiliated Heji Hospital, Changzhi, China

2 Institute of Chinese Materia Medica, Shanghai University of Traditional Chinese Medicine, Shanghai, China

*Address all correspondence to: wanqingzhong3@gmail.com

References

[1] Consensus development conference. Diagnosis, prophylaxis, and treatment of osteoporosis. The American Journal of Medicine. 1993;**94**(6):646-650

[2] Binkley N et al. Osteoporosis in crisis: It's time to focus on fracture. Journal of Bone and Mineral Research. 2017;**32**(7):1391-1394

[3] Lane NE. Epidemiology, etiology, and diagnosis of osteoporosis. American Journal of Obstetrics and Gynecology. 2006;**194**(2 Suppl):S3-S11

[4] Kanis JA et al. Correction to: European guidance for the diagnosis and management of osteoporosis in postmenopausal women. Osteoporosis International. 2020;**31**(4):801

[5] Salari N et al. The global prevalence of osteoporosis in the world: A comprehensive systematic review and meta-analysis. Journal of Orthopaedic Surgery and Research. 2021;**16**(1):609

[6] Lyu H et al. Comparison of Denosumab and bisphosphonates in patients with osteoporosis: A meta-analysis of randomized controlled trials. The Journal of Clinical Endocrinology and Metabolism. 2019;**104**(5):1753-1765

[7] Dimai HP. Use of dual-energy X-ray absorptiometry (DXA) for diagnosis and fracture risk assessment; WHO-criteria, T- and Z-score, and reference databases. Bone. 2017;**104**:39-43

[8] Nam SW et al. The usefulness of trabecular bone score in patients with ankylosing spondylitis. The Korean Journal of Internal Medicine. 2021;**36**(5):1211-1220

[9] Krohn K et al. Dual-energy X-ray absorptiometry monitoring with trabecular bone score: 2019 ISCD official position. Journal of Clinical Densitometry. 2019;**22**(4):501-505

[10] Imai K. Recent methods for assessing osteoporosis and fracture risk. Recent Pat Endocr Metab Immune Drug Discov. 2014;**8**(1):48-59

[11] Grassi L et al. Prediction of femoral strength using 3D finite element models reconstructed from DXA images: Validation against experiments. Biomechanics and Modeling in Mechanobiology. 2017;**16**(3):989-1000

[12] Siddique N et al. Statistical analysis of fat and muscle mass in osteoporosis in elderly population using total body DXA scans. Irish Journal of Medical Science. 2020;**189**(3):1105-1113

[13] Farzi M et al. Quantitating the effect of prosthesis design on femoral remodeling using high-resolution region-free densitometric analysis (DXA-RFA). Journal of Orthopaedic Research. 2017;**35**(10):2203-2210

[14] Hans D, Baim S. Quantitative ultrasound (QUS) in the Management of Osteoporosis and Assessment of fracture risk. Journal of Clinical Densitometry. 2017;**20**(3):322-333

[15] Langton CM, Njeh CF. The measurement of broadband ultrasonic attenuation in cancellous bone--a review of the science and technology. IEEE Transactions on Ultrasonics, Ferroelectrics, and Frequency Control. 2008;**55**(7):1546-1554

[16] Njeh CF et al. Comparison of six calcaneal quantitative ultrasound devices: Precision and hip fracture discrimination. Osteoporosis International. 2000;**11**(12):1051-1062

[17] Karjalainen JP et al. New method for point-of-care osteoporosis screening and diagnostics. Osteoporosis International. 2016;**27**(3):971-977

[18] Casciaro S et al. An advanced quantitative Echosound methodology for femoral neck densitometry. Ultrasound in Medicine & Biology. 2016;**42**(6):1337-1356

[19] Di Paola M et al. Radiofrequency echographic multispectrometry compared with dual X-ray absorptiometry for osteoporosis diagnosis on lumbar spine and femoral neck. Osteoporosis International. 2019;**30**(2):391-402

[20] Fu Y et al. Fragility fracture discriminative ability of radius quantitative ultrasound: A systematic review and meta-analysis. Osteoporosis International. 2021;**32**(1):23-38

[21] Biver E et al. Associations between radius low-frequency axial ultrasound velocity and bone fragility in elderly men and women. Osteoporosis International. 2019;**30**(2):411-421

[22] Zagórski P et al. Does quantitative ultrasound at the calcaneus predict an osteoporosis diagnosis in postmenopausal women from the Silesia Osteo active study? Ultrasound in Medicine & Biology. 2021;**47**(3): 527-534

[23] Adami G et al. Radiofrequency echographic multi spectrometry for the prediction of incident fragility fractures: A 5-year follow-up study. Bone. 2020;**134**:115297

[24] Chan MY et al. Absolute fracture-risk prediction by a combination of calcaneal quantitative ultrasound and bone mineral density. Calcified Tissue International. 2012;**90**(2):128-136

[25] Cortet B et al. Radiofrequency Echographic multi spectrometry (REMS) for the diagnosis of osteoporosis in a European multicenter clinical context. Bone. 2021;**143**:115786

[26] Shuhart CR et al. Executive summary of the 2019 ISCD position development conference on monitoring treatment, DXA cross-calibration and least significant change, spinal cord injury, Peri-prosthetic and orthopedic bone health, transgender medicine, and pediatrics. Journal of Clinical Densitometry. 2019;**22**(4):453-471

[27] Caffarelli C et al. Could radiofrequency echographic multispectrometry (REMS) overcome the overestimation in BMD by dual-energy X-ray absorptiometry (DXA) at the lumbar spine? BMC Musculoskeletal Disorders. 2022;**23**(1):469

[28] Dheeraj D et al. Comparison of quantitative computed tomography and dual X-ray absorptiometry: Osteoporosis detection rates in diabetic patients. Cureus. 2022;**14**(3):e23131

[29] Engelke K et al. Clinical use of quantitative computed tomography and peripheral quantitative computed tomography in the management of osteoporosis in adults: The 2007 ISCD official positions. Journal of Clinical Densitometry. 2008;**11**(1):123-162

[30] Agarwal S et al. Spine volumetric BMD and strength in premenopausal idiopathic osteoporosis: Effect of Teriparatide followed by Denosumab. The Journal of Clinical Endocrinology and Metabolism. 2022;**107**(7):e2690-e2701

[31] Engelke K. Quantitative computed tomography-current status and new developments. Journal of Clinical Densitometry. 2017;**20**(3):309-321

[32] Sfeir JG et al. Evaluation of cross-sectional and longitudinal changes in volumetric bone mineral density in postmenopausal women using single- versus dual-energy quantitative computed tomography. Bone. 2018;**112**:145-152

[33] Wu Y et al. Application of low-tube current with iterative model reconstruction on Philips brilliance iCT elite FHD in the accuracy of spinal QCT using a European spine phantom. Quantitative Imaging in Medicine and Surgery. 2018;**8**(1):32-38

[34] Cheng X et al. Validation of quantitative computed tomography-derived areal bone mineral density with dual energy X-ray absorptiometry in an elderly Chinese population. Chinese Medical Journal. 2014;**127**(8):1445-1449

[35] Kung AW et al. International Society for Clinical Densitometry official positions: Asia-Pacific region consensus. Journal of Clinical Densitometry. 2010;**13**(4):346-351

[36] Cheng X et al. Chinese expert consensus on the diagnosis of osteoporosis by imaging and bone mineral density. Quantitative Imaging in Medicine and Surgery. 2020;**10**(10):2066-2077

[37] Krug R et al. High-resolution imaging techniques for the assessment of osteoporosis. Radiologic Clinics of North America. 2010;**48**(3):601-621

[38] Bandirali M et al. Dose absorption in lumbar and femoral dual energy X-ray absorptiometry examinations using three different scan modalities: An anthropomorphic phantom study. Journal of Clinical Densitometry. 2013;**16**(3):279-282

[39] Agarwal S et al. In vivo assessment of bone structure and estimated bone strength by first- and second-generation HR-pQCT. Osteoporosis International. 2016;**27**(10):2955-2966

[40] Samelson EJ et al. Cortical and trabecular bone microarchitecture as an independent predictor of incident fracture risk in older women and men in the bone microarchitecture international consortium (BoMIC): A prospective study. The Lancet Diabetes and Endocrinology. 2019;**7**(1):34-43

[41] Cataño Jimenez S et al. Dual-energy estimates of volumetric bone mineral densities in the lumbar spine using quantitative computed tomography better correlate with fracture properties when compared to single-energy BMD outcomes. Bone. 2020;**130**:115100

[42] Alvarenga JC et al. Age-related reference curves of volumetric bone density, structure, and biomechanical parameters adjusted for weight and height in a population of healthy women: An HR-pQCT study. Osteoporosis International. 2017;**28**(4):1335-1346

[43] Whittier DE et al. Guidelines for the assessment of bone density and microarchitecture in vivo using high-resolution peripheral quantitative computed tomography. Osteoporosis International. 2020;**31**(9):1607-1627

[44] Liew D et al. Cost-effectiveness of treatment of women aged 70 years and older with both osteopenia and microstructural deterioration. Bone. 2021;**142**:115682

[45] Xu L et al. Validation of goose liver fat measurement by QCT and CSE-MRI with biochemical extraction and pathology as reference. European Radiology. 2018;**28**(5):2003-2012

[46] Cheng X et al. The optimal anatomic site for a single slice to estimate the

total volume of visceral adipose tissue by using the quantitative computed tomography (QCT) in Chinese population. European Journal of Clinical Nutrition. 2018;**72**(11):1567-1575

[47] Lujano-Negrete AY et al. Bone metabolism and osteoporosis during pregnancy and lactation. Archives of Osteoporosis. 2022;**17**(1):36

[48] Chen Y et al. Bone susceptibility mapping with MRI is an alternative and reliable biomarker of osteoporosis in postmenopausal women. European Radiology. 2018;**28**(12):5027-5034

[49] Wang Y et al. Systematic review and meta-analysis: The value of MRI chemical-shift imaging in the evaluation of bone quality in patients with osteoporosis. Ann Palliat Med. 2021;**10**(12):12706-12715

[50] Wu HZ et al. Correlation of bone mineral density with MRI T2* values in quantitative analysis of lumbar osteoporosis. Archives of Osteoporosis. 2020;**15**(1):18

[51] Soldati E et al. Assessment of bone microarchitecture in fresh cadaveric human femurs: What could Be the clinical relevance of ultra-high field MRI. Diagnostics (Basel). 2022;**12**(2):439

[52] Link TM et al. High-resolution MRI vs multislice spiral CT: Which technique depicts the trabecular bone structure best? European Radiology. 2003;**13**(4):663-671

[53] Chang G et al. Finite element analysis applied to 3-T MR imaging of proximal femur microarchitecture: Lower bone strength in patients with fragility fractures compared with control subjects. Radiology. 2014;**272**(2):464-474

[54] Robson MD et al. Human imaging of phosphorus in cortical and trabecular bone in vivo. Magnetic Resonance in Medicine. 2004;**51**(5):888-892

[55] Pierce JL et al. Defining osteoblast and adipocyte lineages in the bone marrow. Bone. 2019;**118**:2-7

[56] Momeni M et al. Sensitivity and specificity assessment of DWI and ADC for the diagnosis of osteoporosis in postmenopausal patients. La Radiologia Medica. 2020;**125**(1):68-74

[57] Zhu HL, Ding JP, Qi YJ. Quantitative evaluation of lumbar spine osteoporosis by apparent diffusion coefficient and signal intensity ratio of magnetic resonance diffusion-weighted magnetic resonance imaging. Zhongguo Gu Shang. 2021;**34**(8):743-749

[58] Griffith JF et al. Reduced bone perfusion in osteoporosis: Likely causes in an ovariectomy rat model. Radiology. 2010;**254**(3):739-746

[59] Kahleova H et al. Effect of a low-fat vegan diet on body weight, insulin sensitivity, postprandial metabolism, and Intramyocellular and hepatocellular lipid levels in overweight adults: A randomized clinical trial. JAMA Network Open. 2020;**3**(11):e2025454

[60] Kuo TR, Chen CH. Bone biomarker for the clinical assessment of osteoporosis: Recent developments and future perspectives. Biomarker Research. 2017;**5**:18

[61] Garnero P, Vergnaud P, Hoyle N. Evaluation of a fully automated serum assay for total N-terminal propeptide of type I collagen in postmenopausal osteoporosis. Clinical Chemistry. 2008;**54**(1):188-196

[62] Delmas PD et al. The use of biochemical markers of bone turnover in osteoporosis. Committee of

Scientific Advisors of the international osteoporosis foundation. Osteoporosis International. 2000;**11**(Suppl. 6):S2-S17

[63] Dai Z et al. Bone turnover biomarkers and risk of osteoporotic hip fracture in an Asian population. Bone. 2016;**83**:171-177

[64] Johansson H et al. A meta-analysis of reference markers of bone turnover for prediction of fracture. Calcified Tissue International. 2014;**94**(5):560-567

[65] Rachner TD, Khosla S, Hofbauer LC. Osteoporosis: Now and the future. Lancet. 2011;**377**(9773):1276-1287

[66] Eastell R, Szulc P. Use of bone turnover markers in postmenopausal osteoporosis. The Lancet Diabetes and Endocrinology. 2017;**5**(11):908-923

[67] Jung K, Lein M. Bone turnover markers in serum and urine as diagnostic, prognostic and monitoring biomarkers of bone metastasis. Biochimica et Biophysica Acta. 2014;**1846**(2):425-438

[68] Baim S, Miller PD. Assessing the clinical utility of serum CTX in postmenopausal osteoporosis and its use in predicting risk of osteonecrosis of the jaw. Journal of Bone and Mineral Research. 2009;**24**(4):561-574

[69] Vilaca T, Gossiel F, Eastell R. Bone turnover markers: Use in fracture prediction. Journal of Clinical Densitometry. 2017;**20**(3):346-352

[70] Jain S, Camacho P. Use of bone turnover markers in the management of osteoporosis. Current Opinion in Endocrinology, Diabetes, and Obesity. 2018;**25**(6):366-372

[71] Takeshita S et al. Osteoblast-specific factor 2: Cloning of a putative bone adhesion protein with homology with the insect protein fasciclin I. The Biochemical Journal. 1993;**294**(Pt 1): 271-278

[72] Liu S et al. Periostin regulates osteogenesis of mesenchymal stem cells from ovariectomized rats through actions on the ILK/Akt/GSK-3β Axis. Genetics and Molecular Biology. 2021;**44**(3):e20200461

[73] Kudo A. The structure of the Periostin gene, its transcriptional control and alternative splicing, and protein expression. Advances in Experimental Medicine and Biology. 2019;**1132**:7-20

[74] Kii I. Periostin functions as a scaffold for assembly of extracellular proteins. Advances in Experimental Medicine and Biology. 2019;**1132**:23-32

[75] Li R et al. Association of Serum Periostin Level with classical bone turnover markers and bone mineral density in Shanghai Chinese postmenopausal women with osteoporosis. Int J Gen Med. 2021;**14**:7639-7646

[76] Li J et al. Plasma periostin as a biomarker of osteoporosis in postmenopausal women with type 2 diabetes. Journal of Bone and Mineral Metabolism. 2021;**39**(4):631-638

[77] Rousseau JC et al. Serum periostin is associated with fracture risk in postmenopausal women: A 7-year prospective analysis of the OFELY study. The Journal of Clinical Endocrinology and Metabolism. 2014;**99**(7):2533-2539

[78] Pepe J et al. Higher serum levels of a cathepsin K-generated periostin fragment are associated with fractures in postmenopausal women with primary hyperparathyroidism: A pilot study. Osteoporosis International. 2021;**32**(11):2365-2369

[79] Xiao SM et al. Association of CDX1 binding site of periostin gene with bone mineral density and vertebral fracture risk. Osteoporosis International. 2012;**23**(7):1877-1887

[80] Kim BJ et al. Plasma periostin associates significantly with non-vertebral but not vertebral fractures in postmenopausal women: Clinical evidence for the different effects of periostin depending on the skeletal site. Bone. 2015;**81**:435-441

[81] Yamamoto M, Sugimoto T. Advanced glycation end products, diabetes, and bone strength. Current Osteoporosis Reports. 2016;**14**(6):320-326

[82] Mercer N et al. Regulation of advanced glycation end product (AGE) receptors and apoptosis by AGEs in osteoblast-like cells. Molecular and Cellular Biochemistry. 2007;**306**(1-2):87-94

[83] Tanaka K et al. Effects of high glucose and advanced glycation end products on the expressions of sclerostin and RANKL as well as apoptosis in osteocyte-like MLO-Y4-A2 cells. Biochemical and Biophysical Research Communications. 2015;**461**(2):193-199

[84] Hein G et al. Advanced glycation end-products pentosidine and N epsilon-carboxymethyllysine are elevated in serum of patients with osteoporosis. Rheumatology (Oxford). 2003;**42**(10):1242-1246

[85] Galliera E et al. Evaluation of circulating sRAGE in osteoporosis according to BMI, adipokines and fracture risk: A pilot observational study. Immunity & Ageing. 2017;**14**:13

[86] Lamb LS et al. Advanced glycation end products and esRAGE are associated with bone turnover and incidence of hip fracture in older men. The Journal of Clinical Endocrinology and Metabolism. 2018;**103**(11):4224-4231

[87] Raška I Jr et al. Prevalence and risk factors of osteoporosis in postmenopausal women with type 2 diabetes mellitus. Central European Journal of Public Health. 2017;**25**(1):3-10

[88] Suzuki A, Yabu A, Nakamura H. Advanced glycation end products in musculoskeletal system and disorders. Methods. 2022;**203**:179-186

[89] Blaine J, Dylewski J. Regulation of the actin cytoskeleton in podocytes. Cell. 2020;**9**(7):1700

[90] Chellaiah M et al. Gelsolin deficiency blocks podosome assembly and produces increased bone mass and strength. The Journal of Cell Biology. 2000;**148**(4):665-678

[91] Zhu W et al. Cytosolic proteome profiling of monocytes for male osteoporosis. Osteoporosis International. 2017;**28**(3):1035-1046

[92] Wang WY et al. Plasma gelsolin is associated with hip BMD in Chinese postmenopausal women. PLoS One. 2018;**13**(5):e0197732

[93] Deng FY et al. Is GSN significant for hip BMD in female Caucasians? Bone. 2014;**63**:69-75

[94] Kobayakawa T et al. Denosumab versus romosozumab for postmenopausal osteoporosis treatment. Scientific Reports. 2021;**11**(1):11801

[95] Genetos DC et al. Impaired osteoblast differentiation in annexin A2- and -A5-deficient cells. PLoS One. 2014;**9**(9):e107482

[96] Zhou X et al. Anxa2 attenuates osteoblast growth and is associated with hip BMD and osteoporotic fracture in Chinese elderly. PLoS One. 2018;**13**(3):e0194781

Section 2

Osteoporosis

Chapter 4

Characteristics of Pathogenetic Links in Vascular Remodeling and Bone Tissue Destruction in Postmenopausal Women with Arterial Hypertension

Tatiana Petelina, Ksenia Avdeeva, Svetlana Bykova, Natalia Musikhina, Liydmila Gapon, Elena Gorbatenko and Anastasiya Shcherbinina

Abstract

The role of nonspecific immune inflammatory vascular response as a link in general pathogenetic mechanisms with change in the elastic properties of the arteries and phenomena of destructive bone changes has attracted great attention. We examined 104 patients (mean age 54.03 ± 9.56) who were divided into three groups: healthy women, with arterial hypertension (AH) and osteopenia and with AH and osteoporosis. The immune inflammatory response markers, endothelial dysfunction, and hormonal and mineral-vitamin status were analyzed simultaneously with 24-hour ambulatory blood pressure monitoring, parameters of vascular wall stiffness, and densitometry to clarify the predictors of cardiovascular and degenerative bone changes in postmenopausal women. For patients with AH and osteopenia, significant parameter associated with the risk of osteoporosis was pulse wave velocity; increase of which exceeded 12.05 m/s was associated with increased risk of osteoporosis by 3.8 times. The levels of pro-inflammatory parameters, interleukin (IL) 6 and 8, tumor necrosis factor-α, high-sensitivity C-reactive protein, and parathyroid hormone were increased and the levels of progesterone and IL-10 were decreased. Timely specialized multidirectional studies of biochemical and instrumental parameters (pulse wave velocity and densitometry) can be the basis for the development of personalized prevention and treatment strategy for women to prevent dangerous cardiovascular and bone complications.

Keywords: atherosclerosis, osteoporosis, cardiovascular diseases, immune inflammatory response, hormonal and vitamin and mineral status, T-Score peak

1. Introduction

Atherosclerosis (AS) and osteoporosis (OP) are currently considered chronic non-infectious epidemics of the 21st century. These diseases are age-related but they are associated with both increase in life expectancy and etiopathogenetic relationships. AS and OP have a number of common features and, above all, asymptomatic course at the onset and a high risk of delayed complications; for AS, these are heart attack and stroke, for OP, low-trauma fractures with a rate of 30–40% [1].

Modern medicine finds it extremely important to identify certain relationships and common pathogenetic mechanisms between various diseases in order to develop an integrated and individualized approach to the treatment and prevention of diseases. The results of experimental and clinical studies conducted over the last decade confirm that AS and OP with asymptomatic onset had common pathogenetic links resulting in manifested complications. A relationship was shown between the development of AS and decrease in bone mineral density (BMD), regardless of the age of patients and increased risk of morbidity and mortality due to AS complications in patients with OP [2].

Various factors affecting bone metabolism are involved in the mechanisms of vascular diseases. In order to assess the relationship between OP and cardiovascular diseases (CVD) caused by AS, surrogate markers of these diseases are commonly used, such as parameters of vascular wall stiffness or vascular calcification and BMD. The vascular and bone tissues appeared to have a number of common properties and vascular calcification consists of the same elements as the bone tissue: calcium salts, type I collagen, phosphates, bone morphogenetic protein, etc. It has been suggested that low BMD may be a direct risk factor for AS of the coronary arteries [3].

The relationship between AS and OP is most evident in postmenopausal women. In estrogen deficiency, the ability of endothelial cells to produce nitric oxide, which supports the elasticity of the arteries and has stimulating effects on the osteoblasts, decreases, which results in endothelial dysfunction and bone metabolism disorders [2, 3].

Together with the deficit of sex steroids, negative calcium balance caused by vitamin D deficiency and reduced absorption of calcium in the intestine is of great importance, which ultimately results in secondary hyperparathyroidism and increased bone resorption [4, 5]. Disorders leading to both OP and CVD include increased activity of the sympathetic autonomic nervous system, which, together with endothelial dysfunction, causes disorders of the microcirculation system. The most important mechanism for reducing BMD is deterioration of bone tissue perfusion associated with disorders of the microcirculation system. Microcirculation that determines the value of peripheral vascular resistance, due to the "steal" syndrome, significantly affects the state of perfusion of internal organs, including bone tissue [6].

The role of angiotensin II in the development of CVD is well known. In addition to the vasoconstrictor effects, it has significant pro-inflammatory activity in the vascular wall (stimulating the production of reactive oxidized particles, inflammatory cytokines, and adhesion molecules) and contributes to the formation and progression of AS. Angiotensin II receptors have been identified in the culture of bone tissue cells (osteoblasts and osteoclasts). Angiotensin II promotes the production of the receptor activator of nuclear factor kappa-B ligand (RANKL) by osteoblasts, which leads to additional activation of the osteoclasts and increased bone resorption, as well as inhibition of bone mineralization [7, 8].

The results of clinical studies of the relationships between BMD and arterial hypertension (AH) and blood pressure (BP) levels have been controversial. Some of them showed a negative relationship between BP and bone density, while others showed no relationship between BP and BMD [9, 10]. There are also published data showing that arterial stiffness is higher in women with moderate cardiovascular risk and postmenopausal OP and is closely associated with BMD and bone turnover markers. It was shown that decrease in BMD of the femoral neck is an independent factor for increase in arterial stiffness. The data obtained allowed us to assume that bone mineral metabolism disorders may be an additional risk factor for vascular wall damage, which must be taken into account when determining patients' total cardiovascular risk [11–14].

With steady aging of the population in the 21st century, data on the association of the processes of cardiovascular remodeling and bone tissue resorption in postmenopausal period remain of interest. More and more attention has been recently paid to the role of nonspecific immune inflammatory vascular response as a link in the common pathogenetic mechanisms of atherosclerotic lesions of the vascular bed with changes in the elastic properties of the arteries and the phenomena of degenerative bone changes, which is of great importance at subclinical level, to provide comprehensive measures for the prevention of complications of these comorbid conditions in general.

The purpose of our work was to study the role of nonspecific immune-inflammatory markers, parathyroid hormone, and female sex hormones as predictors of cardiovascular and degenerative bone changes in postmenopausal women with AH and OP.

2. Materials and methods

The study involved 104 patients (mean age 54.03 ± 9.56 years) who were divided into three groups. Group 1 included 39 healthy women, group 2 – 30 patients with AH and osteopenia, and group 3 – 35 women with AH and OP. The study protocol was approved by the Ethics Committee of Tyumen Cardiologic Research Center, Tomsk National Research Medical Center of the Russian Academy of Sciences, Tomsk, Russia. Before enrollment, each study participant gave a written informed consent to use the study results for scientific purposes.

Exclusion criteria included: presence of acute cerebrovascular accident less than 6 months ago, coronary artery disease, type 2 diabetes mellitus, chronic heart failure of functional class (FC) III-IV (according to the New York Heart Association classification–NYHA), cancer, and mental illness. AH was diagnosed according to the current recommendations of the European Society of Cardiology and Russian Society of Cardiology. The scope of diagnostic measures included: clinical examination, laboratory, and instrumental methods to evaluate the cardiovascular and skeletal systems. The study of the parameters of 24-hour ambulatory blood pressure monitoring (ABPM) was carried out for all examined patients according to the standard scheme, using the oscillometric method, on the equipment of BPLAB LLC "Petr Telegin" (Russian Federation), with the study of standard parameters.

The study of the elastic properties of the vascular walls was carried out using a sphygmograph Vasera VS-1000 Series (Fukuda Denishi, Japan), with the assessment of the following parameters: pulse wave velocity for the elastic-type arteries on the right or left (PWV-R/L) and ankle-brachial index (ABI-R, ABI-L) as a parameter of peripheral vascular blood flow and a screening parameter for the presence of AS of the vessels of the lower extremities. Osteodensitometry was performed using the Siemens Somatom

Emotion spiral computed tomograph. Calcium content CA-HA and the standard deviation of the T-Score peak were assessed (standard values: from 2.0 to −1.0, normal values: from −1.0 to −2.5, and osteopenia: from −2.5 and lower - OP).

Ultrasound scanning of the brachiocephalic arteries was performed; the parameters of intima-media thickness (IMT) of the carotid artery, state of the vascular wall, and the presence of atherosclerotic plaques were taken into account. IMT was determined at a distance of 2 cm from the bifurcation of the common carotid artery on the posterior wall (normal – less than 0.8 mm, the upper limit of normal was 0.9 mm, the thickening was more than 0.9 mm). Atherosclerotic plaque was a local thickening of the arterial wall exceeding 50% or more of the thickness of the adjacent unchanged IMT, protruding into the lumen of the vessel and having different structure compared to unchanged arterial wall and/or thickening of the IMT of more than 1.3 mm [13]. Fasting venous blood was collected into the Vacuette disposable tubes (Japan); the blood was centrifuged for 15 min at 2500 rpm in the Sigma centrifuge (Germany). Patients' blood serum was aliquoted for further freezing (at −70°C).

The parameters of lipid metabolism were studied using the Cobas Integra 400 plus automatic biochemical analyzer (Switzerland). Total cholesterol and triglycerides (TG) in blood serum were determined using the enzymatic colorimetric method; high-density lipoprotein (HDL) and low-density lipoprotein (LDL) were determined using the direct enzymatic colorimetric method; concentrations of apolipoprotein A-I (Apo A-I), apolipoprotein B (Apo-B), and lipoprotein a (Lp(a)) were obtained using immunoturbidimetry with the analytical kits and control materials by Roche Diagnostics Gmb (Germany).

The following biochemical markers of inflammation were determined: high-sensitivity C-reactive protein (hs-CRP) by immunoturbidimetric method using the "C-reactive protein hs" analytical kit (BioSystem, Spain) on the Clima MC-15 semi-automatic, open type analyzer (Spain); interleukin-1β (IL-1β), interleukin-6 (IL-6), interleukin-8 (IL-8), tumor necrosis factor-α (TNF-α) –"sandwich" and homocysteine (HYC) by competitive methods (solid-phase chemiluminescence enzyme immunoassay) using the following analytical kits: IL-1β, IL-6, IL-8, TNF-α, HYC, respectively. The level of sex hormones (estradiol, progesterone, testosterone) and parathyroid hormone were determined using a competitive, solid-phase, chemiluminescent ELISA method with the Siemens Diagnostics reagents; myeloperoxidase was investigated by the sandwich ELISA method using the eBioscience reagents.

Carbohydrate metabolism was assessed based on glucose and glycated hemoglobin (HbA1c) concentrations. Blood glucose was determined by the hexokinase method using the Cobas Integra 400 plus biochemical analyzer. Glycated hemoglobin was determined by chromatography using the Bio-Rad D10 analyzer, USA. The atherogenic coefficient (AC) was calculated as = Apo B/Apo A-I. The parameters of functional activity of endothelium in blood serum: nitrite levels were determined using the Humalyzer 2000 Human biochemical analyzer (Germany) and endothelin-1-21 using the Dynatech semi-automated immunoassay analyzer (Germany).

Statistical data processing was carried out using the Statistica software package (SPSS Inc., ver 11.5). Parameter distribution was tested using the Kolmogorov-Smirnov test. To determine the statistical significance of the differences in continuous values depending on the distribution parameters, one-factor analysis of variance with the Holm-Sidak correction for multiple comparisons or the Kruskal–Wallis criterion with the Bonferroni correction for multiple comparisons were used. Continuous variables represented as M ± SD (mean ± standard deviation) or Me [Q25; Q75] (median and interquartile range). To assess the differences in qualitative variables, the chi-square criterion and the Fisher exact criterion were used. Spearman and Pearson linear correlation coefficient, logistic regression method, and discriminant analysis were used to identify the relationship between the variables.

3. Results and discussion

Characteristics of clinical and anamnestic data of examined patients are presented in **Table 1**. The data presented in the table show that the age of the patients in Groups 2 and 3 significantly differed from Group 1 ($p < 0.005$; $p < 0.001$, respectively). There were no significant differences in smoking and body mass index (BMI) in all groups, or duration of AH in the groups with AH. The percentage of family history of AH in the groups with AH did not differ, but was significantly lower in the control group of healthy patients. Regarding the grade of AH, the maximum percentage of patients with grade 1 AH was in Group 2 and the maximum percentage of patients with grade 3 AH was in Group 3. In addition, patients in group with AH and OP had significantly longer postmenopausal period compared to the 1st group of patients ($p < 0.001$).

According to ABPM, a significant difference in the parameters was observed for the levels of 24-hour systolic BP (SBP 24) and diastolic BP (DBP 24) between healthy subjects in Group 1 and patients in Group 2, as well as for the levels of SBP 24 variability and nighttime DBP between patients in Groups 2 and 3 ($p < 0.01$). The absence of other significant changes in ABPM parameters can be explained by sufficient adherence of patients with AH to antihypertensive therapy. The characteristics of structural

Parameter		Group 1 Healthy patients (n = 39)	Group 2 Patients with AH and osteopenia (n = 30)	Group 3 Patients with AH and osteoporosis (n = 35)	p (groups 1 and 3)
Age (years)		42.92 ± 13.41	58.91 ± 8.28***	62.68 ± 7.16***	<0.001
Smoking		0%	1 (3.4%)	3 (8.6%)	0.181
Non-smoking		39 (100%)	29 (96.6%)	32 (91.4%)	
AH grade	1		11 (36.7%)	5 (14.3%)	0.029
	2		11 (36.7%)	17 (48.6%)	0.386
	3		8(26.6%)	13(37.1%)	0.259
BMI (kg/m²)		25 ± 0.8	26.14 ± 2.48	25.48 ± 2.61	0.460
Waist volume (cm)		71.01 ± 6.08	83.83 ± 8.31	81.65 ± 12.38	0.165
Hips volume (cm)		93.01 ± 1.41	96.22 ± 7.28	97.51 ± 9.93	0.739
Family history of AH	Yes	17 (43.5%)	20 (66.7%)	29 (82.9%)	0.082
	No	22 (56.6%)	10 (33.3%)	6 (17.1%)	
Postmenopausal period (years)		1.0 [1.0;1.75]	7.0 [4.0;10.0]***	10.0 [5.5;21.5]##	<0.001

****p < 0.001 – comparison between groups 1 and 2.*
##p < 0.01 – comparison between groups 2 and 3.
p – comparison between groups 1 and 3.

Table 1.
Clinical and anamnestic characteristics in groups of examined patients.

Parameter	Healthy patients	Patients with AH and osteopenia	Patients with AH and osteoporosis	p (groups 1 and 3)
	(n = 39)	(n = 30)	(n = 35)	
PWV-R, m/s	11.29 ± 0.84	12.99 ± 1.52**	14.82 ± 2.81#	<0.001
PWV-L, m/s	9.60 ± 0.77	13.32 ± 1.44**	15.09 ± 2.97#	0.001
IMT CCA d, mm	0.70 [0.55; 0.80]	0.80 [0.75; 1.0]	0.90 [0.80; 0.90]	0.046
IMT CCA s, mm	0.70 [0.55; 0.75]	0.80 [0.70; 0.85]	0.90 [0.9; 1.00]	0.007
T Score	−0.40 ± 0.22	−1.47 ± 0.93**	−3.08 ± 0.64##	<0.001
AC	—	110.56 ± 16.27	60.37 ± 26.79##	—

***p < 0.01.*
#p < 0.05 – comparison between groups 2 and 3.
##p < 0.01 – comparison between groups 2 and 3.
p – comparison between groups 1 and 3.

Table 2.
Structural and functional characteristics of the vascular wall and bone tissue in the groups of examined patients (M ± SD).

and functional parameters of vascular wall and bone tissue in the groups of examined patients are presented in **Table 2**.

According to the results presented in **Table 2**, PWV-R/L is significantly higher in the groups of patients with AH compared to the control group. The maximum values were registered in Group 3, which significantly exceeded parameters in Groups 1 and 2, which is compliant with the data of other researchers who registered increase in the rigidity of the vascular wall in postmenopausal women [2, 3].

IMT of the common carotid artery (IMT CCA) d/s maximum values were observed in Group 3 of patients with AH and OP, which significantly exceeded the values in Group 1. T-Score and AC were naturally significantly reduced in Group 3 of patients with AH and OP compared to Groups 1 and 2. Correlation analysis of parameters presented in the table showed moderate relationships in Group 2 – PWV-R with IMT CCA d ($r = 0.415$, $p < 0.06$); in Group 3 – AC with PWV-R ($r = 0.871$, $p < 0.06$) and IMT CCA d ($r = -0.673$, $p < 0.002$).

We decided to study common relationships between the studied parameters by determining their relationships with the biochemical parameters of the lipid profile, inflammatory response, and endothelial dysfunction of the vascular wall, as well as parameters of hormonal and mineral-vitamin metabolism.

Laboratory biochemical parameters in the examined groups of patients are presented in **Table 3**.

According to the table, there is a persistent tendency to increase in the levels of total cholesterol and its atherogenic fractions in the groups with AH compared to the control group of patients. There is a clear tendency to increase in myeloperoxidase as a parameter reflecting increased peroxidation process in Group 3 of patients with AH and OP compared to Groups 1 and 2.

According to the results of vascular inflammatory response markers, the levels of hs-CRP, HYC, and IL-8 were significantly higher in Group 3. A tendency to increase in the levels of endothelin-1 and significant increase in nitrites in Group 3 indicate significant endothelial dysfunction in patients with AH and OP. Results in the study of the parameters of hormonal and mineral-vitamin metabolism showed the maximum

Parameter	Healthy patients (n = 39)	Patients with AH and osteopenia (n = 30)	Patients with AH and osteoporosis (n = 35)	p (groups 1 and 3)
Lipid profile				
TCh (mmol/L)	5.09 ± 1.01	5.54 ± 1.15	5.57 ± 1.18	0.124
HDL (mmol/L)	1.59 ± 0.39	1.54 ± 0.4	1.73 ± 0.52	0.230
LDL (mmol/L)	2.83 ± 0.76	3.32 ± 1.05	3.25 ± 1.17	0.081
TG (mmol/L)	1.21 ± 0.75	1.44 ± 0.85	1.18 ± 0.43	0.246
Apo-A (mg/dL)	187.26 ± 35.7	177.13 ± 27.64	178.7 ± 29.0	0.358
Apo-B (mg/dL)	116.02 ± 139.8	105.75 ± 29.78	105.97 ± 29.42	0.261
Apo-A1/Apo-B (mg/dL)	0.53 ± 0.14	0.6 ± 0.19	0.61 ± 0.16	0.112
Myeloperoxidase (mg/dL)	7.69 [2.12; 11.87]	9.75 [7.13; 13.02]	10.07 [8.09; 13.7]	0.05
Inflammatory markers				
High-sensitivity C-reactive protein (hs-CRP) (mg/L)	1.05 [0.45; 3.13]	2.10* [1.01; 4.05]	3.11 [1.76; 5.51]	0.002
TNF-α (pg/ml)	4.47 ± 0.06	4.78 ± 1.35	5.04 ± 1.26	0.219
Homocysteine (μmol/L)	10.88 ± 2.28	12.42 ± 5.47	13.21 ± 5.21	0.076
IL-1β (pg/ml)	2.49 ± 0.46	2.54 ± 0.53	2.67 ± 1.04	0.539
IL-6 (pg/ml)	1.65 ± 0.41	2.05 ± 0.80	1.92 ± 0.67	0.285
IL-8 (pg/ml)	10.40 ± 4.27	10.42 ± 4.64	12.91 ± 4.75#	0.034
IL-10 (pg/ml)	3.71 ± 0.88	3.33 ± 0.81	3.17 ± 0.76*	0.020
Endothelial dysfunction				
Endothelin-1 (fmol/L)	0.47 [0.13; 1.13]	0.32 [0.05; 0.99]	0.51 [0.24; 1.166]	0.269
Nitrites (μmol/L)	56.21 ± 31.13	68.36 ± 33.63	78.25 ± 40.23	0.036
Nitrates (μmol/L)	67.53 ± 38.22	77.61 ± 30.80	90.9 ± 40.33	0.037
Nitrites / Nitrates	34.11 ± 30.31	29.12 ± 22.53	31.5 ± 26.49	0.782
Parameters of hormonal and calcium metabolism, vitamin D				
Estrogen (nmol/L)	35.54 ± 22.34	26.76 ± 9.54	26.7 ± 9.68	0.449
Progesterone (nmol/L)	1.99 [0.99; 4.9]	0.64 [0.64; 0.81]***	0.64 [0.64; 0.82]	<0.001
Testosterone (nmol/L)	0.70 [0.69; 1.01]	0.69 [0.69; 0.72]	0.69 [0.68; 0.69]	0.010
Parathyroid hormone (pg/mL)	22.5 [16.7; 39.5]	29.3 [21.1; 45.9]	37.7 [22.7; 59.0]	0.029
Calcitonin (pg/mL)	1.39 [1.31; 1.61]	1.26 [1.14; 1.56]	1.18 [1.13; 1.49]	0.152
Vitamin D (ng/mL)	46.24 ± 14.82	41.51 ± 21.96*	39.1 ± 15.14	0.01
Total calcium (mmol/L)	2.39 ± 0.11	2.44 ± 0.09	2.35 ± 0.18#	0.226
Ionized calcium (mmol/L)	1.14 ± 0.02	1.17 ± 0.03	1.17 ± 0.43	0.728

$^*p < 0.05$.
$^{***}p < 0.001$ – icomparison between groups 1 and 2.
$^{\#}p < 0.05$ – comparison between groups 2 and 3.
p – comparison between groups 1 and 3.

Table 3.
Characteristics of biochemical parameters in the examined groups of patients (M ± SD).

levels of parathyroid hormone ($p < 0.029$) and decreased levels of estrogens, progesterone, testosterone, a tendency to decrease in calcitonin, total and ionized calcium and significantly low levels of vitamin D ($p < 0.001$) in Group 3 of patients. Our results are consistent with published data and reflect the severity of changes in the parameters of biochemical markers in different groups of patients [2, 3].

The study showed multiple multidirectional moderate correlations ($r = 0.452$, $p < 0.05$) between the presented structural, functional, and biochemical parameters; in Group 3, negative correlations were observed between the peak of T-Score and age, PWV-L/R, 24-hour and night SBP and DBP, duration of menopause, IL-6, hs-CRP and HYC, as well as between PWV-L and estradiol; positive correlations between T-Score and progesterone and between PWV-R/L and IL-6, LDL-cholesterol, hs-CRP, TNF-α, endothelin-1, mean 24-hour SBP, in daily variability of SBP and DBP.

Back in 1935, Allen et al. showed that estrogens dilate blood vessels, improve blood circulation, and normalize cardiac function, and in 1957, Popovici et al. stated that a decrease in estrogen levels resulted in decrease in acetylcholine, which in turn resulted in coronary and arterial ischemic syndrome. Modern data convincingly prove the existence of the relationships between the levels of sex hormones both with CVD and bone destructive processes, and prolonged vascular inflammatory response is considered a pathogenetically associated link in this relationship.

The risks of the development and progression of destructive changes were calculated using logistic regression for the group of AH with osteopenia and OP in postmenopausal period. Thus, for the patients with AH and osteopenia, statistically significant parameter associated with the risk of OP was PWV-R index, which increase by 1 point was associated with 3.8-time increase in the risk of OP (odds ratio (OR) 3.8, 95% confidence interval (CI) 1.81–7.97). There were no reliable relationships between biochemical parameters and the risk of OP at this stage of the study in this group.

In the group of AH with OP, risks of progression of the bone destructive process were observed with changes in certain biochemical markers. Thus, the risk of OP increased: by 2.5 times with increase in IL-6 by 1 pg./mL (OR 1.037 CI 1.01; 1.065, $p = 0.048$), by 2 times with increase in TNF-α by 1 pg./mL (OR 1.99 CI 1.107; 0.58, $p = 0.022$), by 6.5% with decrease in estrogen by 1 nmol/L (OR 0.967 CI 0.935; 1.00, $p = 0.052$), by 18% with increase in HYC by 1 μmol/L (OR 1.18 CI 1.023; 1.361, $p = 0.023$), by 65% with decrease in progesterone by 1 nmol/L (OR 0.348 95% CI 0.164; 0.739, $p = 0.006$), by 3.7% with increase in parathyroid hormone by 1 pg./mL (OR 1.037 95% CI 1.01; 1.065, $p = 0.009$), by 13.6% with increase in IL-8 by 1 pg./ml (OR 1.136 95% CI 1.016; 61.27, $p = 0.025$), by 54% with decrease in IL-10 by 1 pg./mL (OR 0.459 95% CI 0.252; 0.837, $p = 0.011$). As for functional parameters, the risk of OP increased by 6 times with increase in PWV by 1 m/s (OR 6.06 95% CI 2.203; 16.69, $p = 0.00048$).

The characteristics of OR that maximally determine the risk of OP in the groups of AH with osteopenia and OP are presented in **Figures 1** and **2**. According to the data presented in **Figure 1**, in the group of OP, the most reliable parameter that determined the risk of OP was PWV-R value. According to the data presented in **Figure 2**, increased levels of parathyroid hormone and inflammatory markers IL-6 and 8, TNF-α, hs-CRP, as well as decreased levels of progesterone and anti-inflammatory IL-10, were most actively involved in the aggravation of pre-existing destruction of bone tissue.

In addition, during the receiver operating characteristic (ROC) analysis in the group of patients with AH and OP, cut-off points for increased risk of OP progression were determined; for example, with decrease in progesterone levels below 0.93 nmol/L, the risk of OP increases by 9 times (sensitivity 76.9%, specificity 85.7);

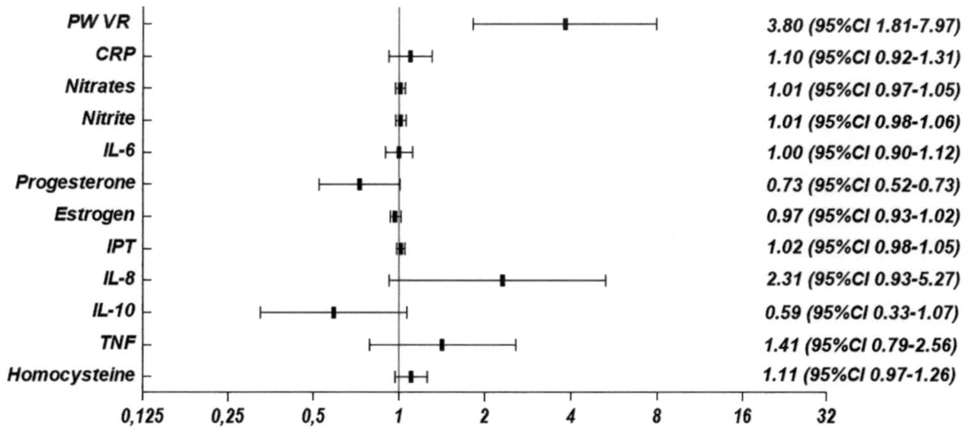

Figure 1.
The odds ratios that maximally determine the risk of osteoporosis in hypertensive patients with osteopenia.

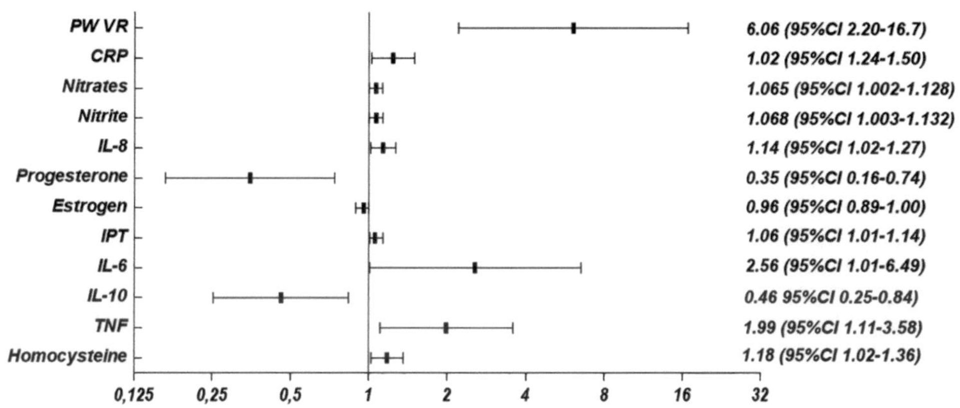

Figure 2.
The odds ratios that maximize the risk of osteoporosis progression in hypertensive patients with pre-existing osteoporosis.

with increase in parathyroid hormone levels over 28.14 pg./mL, the risk of OP increases by 3.7% (sensitivity 68.6%, specificity 69.2%); with increase in IL-8 over 10.25 pg./mL, the risk of OP increases by 13.6% (sensitivity 71.4%, specificity 64.1); with decrease in IL-10 levels below 3.465 pg./mL, the risk of OP increases by 54.1% (sensitivity 66.7%, specificity 62.9%). As for functional parameters, with increase in PWV-R index over 12.05 m/s, the risk of OP increases by 6 times (sensitivity 87.1%, specificity 89.3).

The results of the discriminant analysis clarified that the most significant parameters for the progression of OP in Group 3 of patients were progesterone, parathyroid hormone, IL-8, and PWV-R. The model obtained as a result of the discriminant function calculations is statistically significant (the Wilks' lambda is 0.201, $p < 0.001$) with a canonical correlation coefficient of 0.894.

At the same time, the greatest diagnostic contribution to the progression of OP is made by the level of progesterone (standardized coefficient 0.843). The standardized coefficients of parathyroid hormone (0.523), IL-8 (0.367) and PWV-RC (0.413) indicate approximately the same diagnostic significance of these variables. The equation of the resulting discriminant function is as follows: F = −2.618 + 42.951* progesterone +1.293* parathyroid hormone +0.749* IL-8 + 1.025* PWV-R.

The specificity of this model was 100%, the sensitivity was 92%; 96% of the original observations were classified correctly. To enable classification, centroids were calculated for each group and a cut-off point that allows for more accurate identification of group membership. The mean value of the function for Group 1 was −1.915, for Group 2 - 1.992; the cut-off point (or threshold value) was equal to the function value of 0.039.

4. Conclusion

The demographic situation worldwide is characterized by a steady increase in the number of elderly people. With steady aging of the population, the problem of increased number of socially significant diseases in women is gaining increased interest. The most common causes of disability and mortality in older postmenopausal women include clinical consequences of AS and OP: cardiovascular accidents and bone fractures. It is known that many factors influencing bone metabolism are involved in the mechanisms of vascular diseases. There is a similarity in the course of these diseases, since they can be asymptomatic for many years and often have clinical manifestations after menopause.

Recently, the role of a nonspecific immune inflammatory vascular response as a link in general pathogenetic mechanisms of atherosclerotic lesions of the vascular bed and phenomenon of destructive bone changes has attracted great attention.

The multimarker approach in the study of common links in the pathogenesis of socially significant diseases enabled the clarification of the main risk factors, laboratory levels of nonspecific immune inflammatory response markers, and parameters of hormonal and vitamin status, which determine a degree of impairment of the elastic properties of the vascular wall and the risk of progression of OP can be predictors of CVD and degenerative bone complications in postmenopausal women with AH.

In our study, the following markers of vascular inflammation were increased: hs-CRP, HYC, IL-8, endothelin-1, parathyroid hormone, total cholesterol, and atherogenic lipid fractions, with a simultaneous decrease in the levels of estrogen, progesterone, calcium, and vitamin D. Multiple regression relationships between inflammatory parameters and the parameters of lipid metabolism and hormonal-vitamin status were observed.

The results of the study indicate early examinations of women with AH to detect increased rigidity of the vascular wall and reduced bone mineral density, which create the conditions for increased risk of development and progression of subclinical AS and OP prior to postmenopause.

Timely in-depth examination of women with AH in premenopause should become the main strategy for the development of a personalized prevention and therapy for women in order to prevent socially significant cardiovascular and bone complications such as coronary artery disease, stroke, and low-trauma fractures.

4.1 Limitations of the study

This is a pilot project, which requires an expansion of the patient sample to clarify the subtle mechanisms of the interrelationship of the processes under study. The planned comparative characteristics of the studied parameters between groups of women and men will require an expansion of the range of statistical methods for processing the studied data.

4.2 Perspectives

It is planned to continue recruiting patients, expanding research methods, in particular, conducting a study on the effect of gut microbiota as an additional risk factor for vascular and degenerative bone complications in postmenopausal patients with arterial hypertension.

Author details

Tatiana Petelina*, Ksenia Avdeeva, Svetlana Bykova, Natalia Musikhina, Liydmila Gapon, Elena Gorbatenko and Anastasiya Shcherbinina
Tyumen Cardiology Research Center, Branch of Tomsk National Research Medical Center of the Russian Academy of Sciences, Tyumen, Russia

*Address all correspondence to: petelina@infarkta.net

References

[1] Skripnikova IA, Abirova ES, Alikhanova NA, Kosmatova OV. Vessel stiffness, calcification and osteoporosis. Common pathogenetic components. Cardiovascular Therapy and Prevention. 2018;**17**(4):95-102. DOI: 10.15829/1728-8800-2018-4-95-102

[2] Alikhanova NA, Skripnikova IA, Tkacheva ON, Abirova ES, Novikov VE, Murashko LM, et al. Association of vessel stiffness parameters and subclinical atherosclerosis and mass of bone tissue in postmenopausal women. Cardiovascular Therapy and Prevention. 2016;**15**(2):51-56. DOI: 10.15829/1728-8800-2016-2-51-56

[3] Skripnikova IA, Alikhanova NA, Tkacheva ON, Abirova ES, Novikov VE, Murashko LM, et al. Bone mineral density and vascular wall condition depending on the status of replicable cellular aging in postmenopausal women. Osteoporosis and Bone Diseases. 2015;**18**(3):13-17. DOI: 10.14341/osteo2015313-17

[4] Povoroznyuk VV, Pludovski P, Balatskaya NI, Muts VY, Klimovitsky FV, Reznichenko NA, et al. Vitamin D deficiency and insufficiency: Epidemiology, diagnosis, prevention and treatment. Kiev: Zaslavsky A.Y. 2015. 262 p. (In Russ.)

[5] Somjen D, Weisman Y, Kohen F, Gayer B, Limor R, Sharon O, et al. 25-hydroxyvitamin D3-1alpha-hydroxylase is expressed in human vascular smooth muscle cells and is upregulated by parathyroid hormone and estrogenic compounds. Circulation. 2005;**111**(13):1666-1671. DOI: 10.1161/01.CIR.0000160353.27927.70

[6] Wang TJ, Pencina MJ, Booth SL, Jacques PF, Ingelsson E, Lanier K, et al. Vitamin D deficiency and risk of cardiovascular disease. Circulation. 2008;**117**(4):503-511. DOI: 10.1161/CIRCULATIONAHA.107.706127

[7] Sagalovsky S. Bone remodeling: Cellular-molecular biology and cytokine RANK-RANK-OPG system and growth factors. Crimea Journal of Experimental and Clinical Medicine. 2013;**3**(1-2):36-43. UDC 616-001.5.616.71-003.93

[8] Heymann MF, Herrison F, Dovaine JM, Charrier C, Battaglia S, Passuti N, et al. Role of the RANK-RANK-OPG triad in calcification of the atheromathous plaques: Comparison between carotid and femoral beds. Citokine. 2012;**58**(2):300-306. DOI: 10.1016/j.cyto.2012.02.004

[9] Datta HK, Ng WF, Walker JA, Tuck SP, Varanasi SS. The cell biology of bone metabolism. Journal of Clinical Pathology. 2008;**61**(5):577-587. DOI: 10.1136/jcp.2007.048868

[10] Tamaki J, Iki M, Hirano Y, Sato Y, Kajita E, Kagamimori S, et al. Low bone mass is associated with carotid atherosclerosis in postmenopausal women: The Japanese population-based osteoporosis (JPOS) Cohort study. Osteoporosis International. 2009;**20**(1):53-60. DOI: 10.1007/s00198-008-0633-z

[11] Kim LN, Jang HM, Kim SK, Ko KD, Hwang IC, Suh HS. Association of arterial stiffness and osteoporosis in healthy men undergoing screening medical examination. Journal of Bone Metabolism. 2014;**21**(2):133-141. DOI: 10.11005/jbm.2014.21.2.133

[12] Li X-S, He H, Zhao Y-L, Li Y, Liu Z-P, Liu T, et al. Bone mineral density

is negatively associated with arterial stiffness in men with hypertension. Journal of Clinical Hypertension. 2016;**18**(11):1106-1111. DOI: 10.1111/jch.12848

[13] Cecelia M, Jiang B, Bevan L, Frost ML, Spector TD, Chowienczyk PJ. Arterial stiffening relates to arterial calcification but not to noncalcified atheroma in women. Journal of the American College of Cardiology. 2011;**57**(13):1480-1486. DOI: 10.1016/j.jacc.2010.09.079

[14] Avramovski P, Avramovska M, Lasaverski M, Sikole A. Femoral neck and spine bone mineral density-surrogate marker of aortic calcification in postmenopausal women. Anatolian Journal of Cardiology. 2016;**16**(3):202-209. DOI: 10.5152/akd.2015.6016

Chapter 5

Non-GCs Drug-Induced Osteoporosis

Hesham Hamoud

Abstract

Medications that cause osteoporosis are numerous and common. While helping to correct one problem, they may be putting you at greater risk of having osteoporosis. A variety of drugs may cause bone loss by lowering sex steroid levels (e.g., aromatase inhibitors used in breast cancer and GnRH agonists used in prostate cancer), interfering with vitamin D levels (liver-inducing antiepileptic drugs), or directly affecting bone cells (chemotherapy, phenytoin, or thiazolidinediones) which divert mesenchymal stem cells from osteoblastogenesis to adipocytogenesis, consequently, an imbalance occurs between bone formation and resorption, as well as between soft organic matrix and hard inorganic matrix. Besides effects on the mineralized matrix, interactions with collagen and other nonmineralized matrix components can decrease bone biomechanical competence without affecting bone mineral density (BMD). Here is a quick narrative for a number of disease medications that can cause osteoporosis if taken for long periods without a preventive program of minerals and vitamins. Rheumatoid arthritis, inflammatory bowel disease, asthma, acid reflux, thyroid dysfunctions, seizures, endometriosis, aromatase inhibitors, hypertension, contraceptive Depo-Provera, antidepressant (SSRIs, SNRIs), glitazones for type 2 DM treatment.

Keywords: amiodarone, proton pump inhibitors, aromatase inhibitors, methotrexate, inflammatory bowel disease, seizures, antidepressant, hypertension, diabetes and antidiabetics, osteoporosis, BMD, risk of fracture

1. Introduction

Medications that cause osteoporosis are numerous and common. While helping to correct one problem, they may be putting you at greater risk of having osteoporosis. Several drugs and drug classes can decrease BMD, including thiazolidinediones, and consequently increase fracture risk; other drugs, such as selective serotonin reuptake inhibitors (SSRI), do not necessarily increase bone loss, but they may increase fracture risk, possibly resulting from an increased risk of falls due to effects on postural balance mediated by central nervous system effects. Amiodarone is a potent antiarrhythmic drug. It is a benzofuran-derived, iodine-rich compound with some structural similarity to thyroxine (T4). Amiodarone contains approximately 37% iodine by weight. Each 200-mg tablet is estimated to contain about 75 mg of organic

IntechOpen

iodide, 8–17% of which is released as free iodide. Thyroid abnormalities have been noted in up to 14–18% of patients receiving long-term amiodarone therapy. 2010 FDA warning: proton pump inhibitors and increased fracture risk revised warning for PPI: possible increased risk of hip, wrist, and spine fractures.

Aromatase inhibitors stop the production of estrogen in postmenopausal women. Aromatase inhibitors work by blocking the enzyme aromatase, which turns the hormone androgen into small amounts of estrogen in the body.

Osteoporosis can arise as a consequence of some rheumatic diseases, as RA itself can contribute to osteoporosis through systemic inflammation; immobility and medications other than glucocorticoids like long-term use of or methotrexate that inhibits osteoblastic differentiation leading to a reduction in bone formation and an increased risk of osteopathy. Patients with IBD are more likely than the general population to experience bone loss due to malnutrition, vitamin D and calcium malabsorption and deficiency, vitamin K insufficiency, immobilization, and underlying inflammatory state.

Long use of such medications leads to decreased bone biomechanical capability and thus a decreased density of bone and an increased risk of fractures.

All patients who have been receiving such medications should undergo a DEXA scan and lateral spine X-ray to check for osteoporosis. People with fragility fractures or at high risk of developing fractures should avoid such medications. Moreover, non-pharmacological measures such as calcium/vitamin D nutrition and exercise should be encouraged. In general, Non-GCs Drug-Induced Osteoporosis is treated with the same medications that are used for general health care when the BMD (T score < 2) or higher.

In this chapter, we aimed to summarize most of these medicines to make them easily accessible for rheumatologists, orthopedists, and anyone else interested in managing osteoporosis (**Figure 1**).

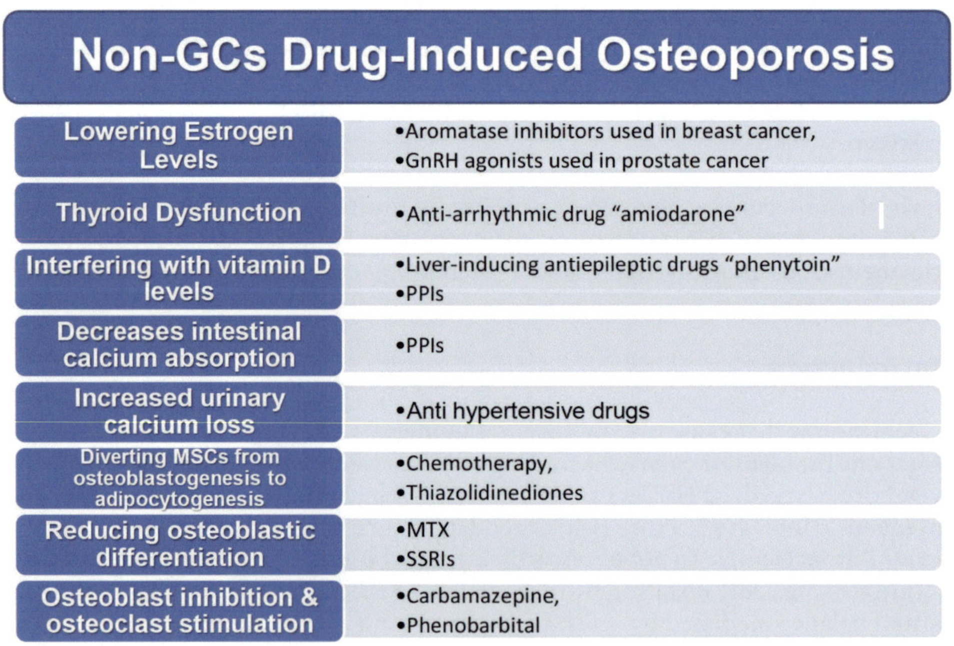

Figure 1.
Non-GC drug-induced osteoporosis: mode of actions with examples.

2. Amiodarone-induced osteoporosis

Due to its high iodine content and direct harmful effect on the thyroid gland, amiodarone is however linked to a range of side effects, including thyroid dysfunction (both hypo- and hyper-thyroidism).

A strong antiarrhythmic medication called amiodarone is used to treat supraventricular and ventricular tachyarrhythmias. It is an iodine-rich molecule produced from benzofurans that resemble thyroxine structurally in several ways (T4). Iodine makes up roughly 37% of the weight of amiodarone. It is believed that each 200-mg tablet contains 75 mg of organic iodide, of which 8–17% is released as free iodide. Standard maintenance therapy uses 200 mg of amiodarone, which is 100 times the recommended daily intake [1, 2].

Up to 14–18% of patients undergoing long-term amiodarone therapy develop thyroid problems which range from aberrant results from thyroid function tests to overt thyroid dysfunction, which could be one of the following: 1- Amiodarone-induced thyrotoxicosis (AIT) or 2- Amiodarone-induced hypothyroidism (AIH) [3, 4]. Both can appear in thyroid glands that appear to be normal or in glands that already have abnormalities.

2.1 Pathophysiology

The thyroid is subject to a variety of impacts from amiodarone.

1. Amiodarone reduces the peripheral conversion of T4 to (T3) and the elimination of both T4 and reverse T3 by inhibiting type 15′-deiodinase enzyme activity (rT3). As a result, the serum levels of T4 and rT3 rise while T3 levels drop by 20–25%.

2. Amiodarone prevents T4 and T3 from entering peripheral tissue. After 1–4 months of amiodarone therapy, serum T4 levels rise by an average of 40% [2, 5, 6].

3. Thyroid-stimulating hormone (TSH) levels rise as a result of feedback regulation's inhibition of type 25′-deiodinase enzyme activity in the pituitary during the first 1–3 months. There is no need for T4 replacement in these patients based on this. In 2–3 months, serum TSH levels return to normal as T4 concentrations sufficiently increase to overcome the gap in T3 synthesis. There may be a diminished response of TSH to thyroid-releasing hormone (TRH) [7].

4. Amiodarone and its metabolites may directly damage thyroid follicular cells, resulting in thyroiditis that is destructive.

5. At the cellular level in the heart, amiodarone and its metabolite desethylamiodarone can function as a competitive antagonist of T3 [7].

While a patient is using amiodarone or even months after stopping the medication, thyrotoxicosis might happen. Once treatment has lasted the first 18 months, hypothyroidism is uncommon [8].

2.2 Two forms of AIT have been described

1. Type 1 usually affects patients with latent or preexisting thyroid disorders and is more common in areas of low iodine intake. Type 1 is caused by iodine-induced excess thyroid hormone synthesis and release.

2. Type 2 affects people whose thyroid glands were previously healthy and is brought on by a destructive thyroiditis that causes the thyroid follicular cells to become destroyed and leak preformed thyroid hormones. However, mixed types of AIT with characteristics of destructive processes and excess iodine may develop in an aberrant thyroid gland [9].

2.2.1 AIT signs and symptoms include the following

1. Unaccounted-for weight loss

2. A greater tolerance for heat or perspiration

3. extreme musculature weakness

4. Unknown exhaustion

5. Emotional brittleness

6. Constant stools

7. Oligomenorrhea

8. Panic, trembling, or palpitations [10]

2.2.2 Symptoms of AIH include the following

1. Fatigue

2. Lassitude

3. Intolerance to cold

4. Mental slowness

5. Weakness

6. Constipation

7. Menorrhagia

8. Dry skin [10]

3. Proton-pump inhibitors and risk of fractures

PPIs are among the most frequently recommended treatments worldwide in clinical practice. Although the majority of patients handle PPIs well overall, there is growing concern over a possible link between PPI use and an increased risk of bone fracture [11, 12]. Indeed, the correlation between PPI medication and the incidence of fracture has been documented in numerous observational studies [13–15]. The findings of each study differ significantly from one another. According to meta-analyses of the evidence, PPI medication is often linked to a higher risk of fracture [16–20]. The Food and Drug Administration (FDA) also issued a safety advisory in May 2010 addressing a potential increase in fracture risk of hip, wrist, and spine fractures associated with PPI usage and recommended that no more than three 14-day treatment sessions should be taken in a year, based on seven epidemiologic studies and claims data base analysis (no randomized trials), while they recognized that additional data were required [18].

3.1 Pathophysiology

PPIs are strong inhibitors of stomach acid secretion, which is thought to be important for calcium absorption by enhancing the solubility of calcium salts that are insoluble leading to decreases intestinal calcium absorption and ultimately causing a decline in bone mineral density [15]. Regarding the impact of PPI use on calcium absorption, there is a paucity of clinical evidence and inconsistent findings. Furthermore, PPI use may induce hypomagnesemia, which could increase the fracture risk, although this is also controversial [17]. Increased fracture risk after 1–7 years of treatment. Risk factors include age > 50, "high dose" and longer duration. Zhou et al. [20] stated that PPI use for less than a year was also linked to an increased risk of hip fracture. This finding may undermine the idea that PPI use increases the risk of fracture through biochemical mechanisms (such as changes in calcium absorption or bone mineral density). Further research is required to elucidate any other pathways that may exist and have an impact on bone mineralization or bone quality directly [19, 20].

4. Aromatase inhibitors (AIs)-induced bone loss

The majority of adverse reactions to aromatase inhibitors (AIs) affect the musculoskeletal system and can be divided into three groups [21]:

1. Metabolic bone disease, which increases the risk of fractures;

2. Arthralgia syndrome; and

3. Autoimmune rheumatic illnesses.

All of these adverse outcomes begin to manifest after varying amounts of time have passed since the start of treatment with AIs. Although the precise pathophysiology is not fully understood, the pathogenetic pathways endorsed to explain these disorders are primarily related to the estrogen deficiency caused by a prolonged AIs treatment [21].

4.1 Pathophysiology of AIs-induced bone loss

The hypoestrogenic state brought on by AIs accelerates bone loss at the areas with high levels of trabecularity (vertebral body) and significantly increases bone resorption. In fact, a lack of estrogen alters the dynamic equilibrium between the osteoblasts and osteoclasts activities. Tumor necrosis factor (TNF) and receptor activator of nuclear factor-kB ligand (RANKL), which serve as the main mediators for osteoclast activation and maturation, are more likely to be secreted by T cells as a result of this situation. The equilibrium between RANKL and osteoprotegerin (OPG), a soluble RANKL decoy receptor that blocks the binding of RANK to RANKL and inhibits the osteoclast activity, actually maintains the proper functioning of osteoblasts and osteoclasts [22–28].

Finally, the pathologic bone remodeling seen following AIs therapy may be caused by genetic variations of the RANK/RANKL/OPG pathway. The RANKL/OPG ratio was shown to be altered as a result of the rs7984870 SNP in the RANKL gene, which had detrimental effects on bone health. Despite the fact that Exemestane (Aromasin) appeared to have a bone-sparing impact in preclinical investigations, which is likely due to its androgenic nature, bone loss was documented for all AIs in clinical trials, which largely evaluated these medication's effectiveness in breast cancer. According to some reports, the rate of bone loss following AIs therapy is two times higher than in postmenopausal women in good health [25]. This data led investigators to propose that several pathophysiological mechanisms, such as those influencing bone geometry, bone microstructure, other aspects of bone quality, may be responsible for bone fragility in women who have received AIs treatment.

5. Rheumatoid disease and methotrexate

Low-dose MTX is regarded as an effective RA treatment since it reduces joint stiffness, pain, and inflammation while also greatly delaying bone deterioration. It is generally known that generalized osteoporosis can arise in RA per se. Three main causes have been proposed as the mechanism of this osteoporosis [29]:

1. Systemic rheumatoid inflammation;

2. Immobility; and

3. Medications like corticosteroids.

Since MTX reduces rheumatic inflammation and permits an increase in physical activity, this medication may help with OP brought on by RA. However, MTX has been shown to have a negative impact on bone among RA patients and animal models. Sally et al. described two cases of MTX osteopathy with fractures in rheumatoid arthritis patients getting long-term low-dose MTX treatment. MTX osteopathy has been mentioned in an increasing number of papers [30]. Even though glucocorticoids prefer cancellous bone, MTX-induced bone loss and fractures mostly affected cortical bone. Patients with rheumatic disease who underwent histological investigation revealed that MTX osteopathy has impaired bone formation, as evidenced by a decreased osteoblast surface and a lower mineral apposition rate. Additionally, May et al. observed that low-dose MTX impairs bone formation and increases bone resorption in both normal and ovariectomized mice, resulting in osteopenia [31, 32]. However, MTX's inhibition of the development of marrow osteoblast precursor cells leaves unclear the particular mechanism by which it reduces bone production. Additionally, MTX significantly reduced ALP activity and prevented calcified nodules from forming in cultures of marrow stromal cells. Given that May et al. found that MTX inhibits matrix mineralization using terminally developed osteoblasts, it is possible that MTX also suppresses mature osteoblasts. The transcription factor Cbfa1 has recently been identified as a key player in osteoblastogenesis. Therefore, a study into how MTX affects osteoprogenitor cells' expression of Cbfa1 may be useful for understanding the molecular basis of MTX osteopathy. Furthermore, bone metabolism is hampered by disease activity [29–34].

6. Inflammatory bowel disease (IBD)

IBD, which is predominantly made up of Crohn's disease (CD) and ulcerative colitis (UC), is linked to a number of systemic problems, including extraintestinal manifestations (EIMs) which are prevalent among 40% of IBD patients. The most well-known (EIMs) are as follows:

1. Liver illnesses (primary sclerosing cholangitis and primary biliary cirrhosis),

2. Articular symptoms,

3. Skin lesions (such as erythema nodosum and pyoderma gangrenosum) [35–37].

Patients with IBD are more likely than the general population to experience bone loss. Osteopenia and osteoporosis are manifested by a decrease in bone mineral density (BMD), caused by chronic inflammation [38, 39]. According to cross-sectional studies, IBD patients have a wide-ranging prevalence of low BMD. Depending on the study population, location, and methodology, the prevalence of osteopenia and osteoporosis can range from 22% to 77% and 17% to 41%, respectively [40].

6.1 Pathophysiology of osteoporosis in inflammatory bowel disease

In addition to corticosteroid use, aging, smoking, malnutrition, vitamin D and calcium malabsorption and deficiency, immobility, and the underlying inflammatory condition are risk factors for osteoporosis in inflammatory bowel disease. According to research by Bernstein et al., patients with IBD have a 40% higher incidence of fractures than the general population [41].

6.2 Inflammation

Numerous factors have a significant impact on bone metabolism, but there is growing evidence that inflammation itself has a role in osteoporosis among patients with IBD. Even without the use of drugs like corticosteroids, some investigations in newly diagnosed patients with IBD showed a decline in BMD. Osteoporosis and a increased fracture incidence are associated with a number of chronic inflammatory diseases [42–44]. The production of pro-inflammatory cytokines including interleukin-1 (IL-1), tumor necrosis factor alpha (TNF-), IL-6, IL-11, IL-15, and IL-17, is linked to increased bone resorption and decreased bone formation. RANK/RANKL/osteoprotegerin is probably the major mechanism implicated in the onset of osteoporosis in IBD and other inflammatory illnesses. In a study of 137 IBD patients, Reffitt et al. found that those with prolonged illness remission had higher BMD [45–47].

6.3 Smoking

The amount and duration of smoking may influence how smoking affects bones. Smoking may affect 25 hydroxylases in the liver, which lowers serum 25-hydroxyvitamin D levels, altering the hepatic metabolism of vitamin D. This may explain why smoking and vitamin D deficiency appear to be related. Also, there is proof that smoking can affects the gastrointestinal absorption of calcium. Although it is debatable whether smoking affects estradiol levels, certain research studies have shown that smoking affects the production and metabolism of estrogen. Smoking accelerates the hepatic metabolism of estradiol and nicotine may diminish the synthesis of estrogen. Also, smokers have greater serum levels of the sex hormone-binding protein, which lowers free estradiol levels. All IBD patients should be encouraged to quit smoking, though, as it lowers the risk of secondary consequences such as heart disease, lung cancer, and changes in bone health [48–50].

6.4 Malnutrition

Nutritional deficits linked to inflammatory bowel disorders have been mentioned as additional pathogenic pathways causing low bone mineral density. There have been reports of calcium insufficiency in Crohn's disease due to either inadequate intake or poor intestinal absorption [51–53].

Patients with ulcerative colitis and/or Crohn's disease have been found to have more vitamin D deficiencies compared to the control healthy population reference range. Elevated levels of bone turnover markers coexist with decreased vitamin D level in Crohn's disease patients compared to controls. In general, patients with inflammatory bowel disease have lower vitamin D status for a number of reasons, including [54]: lack of vitamin D lowers calcium levels and triggers secondary hyperparathyroidism, which in turn promotes osteoclastogenesis, increases bone resorption, and causes osteopenia and osteoporosis [55].

A. Decreased efficiency of intestinal absorption of vitamin D due to ileopathy,

B. Disrupted enterohepatic circulation of vitamin D,

C. Decreased dietary intake,

D. Decreased sun exposure, and

E. Renal insufficiency.

Vitamin K insufficiency may potentially have a role in osteopenia related to IBD. Because of ileopathy, some patients may absorb this fat-soluble vitamin. However, the discrepancies in vitamin K status between patients with ulcerative colitis and Crohn's disease may result from changed bacterial flora that produces less vitamin K. Additionally, it is likely that antibiotics, which are frequently used to treat inflammatory bowel disease patients, could eradicate flora that produces vitamin K [54, 56].

7. Impaired bone health and seizures

Long-term use of anti-seizure medications (ASM) has been linked in numerous studies to the development of osteoporosis, which affects between 11% and 31% of epilepsy patients and increases fracture risk by 2–6 times compared to the general population. The increased risk of fractures in epileptic patients can be attributed to a number of factors, such as fractures brought on by seizures and a higher chance of falling due to both the convulsions themselves and the adverse effect of ASM on balance [57–62]. There are other factors that contribute to the increased fracture risk in individuals with epilepsy, as evidenced by the fact that seizure-related fractures in people with epilepsy only make up 25–43% of all fractures. Comparing enzyme-inducing ASM (EIASM) to non-enzyme-inducing ASM (NEIASM), previous research studies have demonstrated that EIASM has a deleterious impact on bone mass and the onset of osteoporosis. ASM polytherapy has additionally been linked to osteopenia [63–66].

7.1 Pathophysiological mechanisms for increased fracture risk in patients with seizures

ASM has been linked to numerous studies of negative effects on bone strength. Although the precise mechanism is not entirely understood, ASM may impact bone quality and bone mass through a variety of methods. But it is widely accepted that certain medications, particularly the EIASM such as carbamazepine and phenytoin, stimulate the hepatic cytochrome P450 system, leading to a variety of endocrine complications [64, 65]. Among these include altered sex hormone-binding globulin concentrations and sex hormone disturbances, but most frequently, EIASM therapy has been proven to cause increased vitamin D metabolism, low vitamin D levels, impaired calcium absorption, and resultant hypocalcemia. These modifications lead to secondary hyperparathyroidism and osteoclastic bone resorption that is activated by parathyroid hormone (PTH), which causes bone loss and decreased bone mineralization [66–69]. This is in line with the common observation that patients with epilepsy have a tendency to have lower BMD, lower levels of 25-OH vitamin D, and higher levels of alkaline phosphatase and PTH. Additionally, a number of ASM carbamazepine, phenytoin, and phenobarbital have unfavorable direct effects on bone metabolism including osteoblast inhibition and osteoclast stimulation [70–72].

8. Hypertension and osteoporotic fracture

Osteoporosis and hypertension are common and frequently comorbid disorders in the aged population. Among elderly people, hypertension affects 20–40% of people. Similar to hypertension, osteoporosis affects 20–30% of postmenopausal women globally [73, 74].

8.1 Pathophysiological mechanisms for increased fracture risk in patients with hypertension

According to recent studies, both disorders may have the same etiopathology [75]. Additionally, some hypotensive medications may influence bone mineral density and exacerbate osteoporosis. There are several genetic and etiological similarities between osteoporosis and hypertension. Aging, menopause, and physical inactivity are risk factors for both hypertension and osteoporosis. Human and animal studies have shown that elevated blood pressure is linked to aberrant calcium metabolism, which increases urine calcium loss [76–79]. These hypertension-related anomalies may ultimately attribute to increased bone loss and decreased bone mineral density (BMD). The overall cumulative incidence of any fracture, hip fracture, and clinical vertebral fracture for men with hypertension was16.3, 3.3, and 5.7 per 1000 person-years compared with 11.3, 2.8, and 4.5 per 1000 person-years for those without hypertension, respectively [80–84].

In women, additionally, the cumulative total fracture incidence was greater in the hypertensive group compared to the non-hypertensive group (27.6 vs. 21.6 per 1000 person-years for any fracture; 5.7 vs. 1.1 for hip fracture, and 9.3 vs. 8.8 for vertebral fracture) [85].

In contrast to the non-hypertensive group, the cumulative incidence of any hip fracture in women was significantly higher in the hypertensive group [86].

There is a physiologic basis for the association between hypertension and osteoporosis. High blood pressure is linked to increased urinary calcium loss, which impairs the calcium balance necessary for bone remodeling. In fact, an epidemiological study discovered that elevated blood pressure was associated with an increased rate of mineral loss from the bone [87]. Furthermore, high levels of the parathyroid hormone are linked to hypertension and accelerated bone turnover, reducing bone mass, and bone quality. Finally, high blood pressure may gradually harm brain regions involved in balance and gait regulation, which could increase the risk of falls and consequent fractures. These findings imply that appropriately managed blood pressure may potentially promote bone health and protection against fragility fracture given those two closely associated medical problems [88–90].

9. Serotonin reuptake inhibitors and bone health

Anhedonia, insomnia, anorexia, fatigue, and cognitive dysfunction are symptoms of major depressive disorder (MDD). Osteoporosis and fractures are more likely to occur in people with MDD [91–94].

SSRIs are associated with an increased fracture risk compared to nonusers, according to a number of observational studies [95–98]. Comparing 124,655 fracture cases with 373,962 controls, Danish national registers found that the use of SSRIs increased the risk of hip and vertebral fractures [99–102]. A Dutch study found that SSRI use was associated with an early increase in fracture risk, which peaked within 8 months of use but decreased after discontinuation. The risk of fracture among old patients taking SSRIs is almost threefold in Taiwanese case–control studies [103].

In a cross-sectional study of 5995 patients, SSRIs significantly reduced hip BMD (4%) and spine BMD (6%) compared with nonusers. This was further confirmed in a cohort study of nearly 3000 women divided into three groups: SSRI users

(198), tricyclic antidepressants (TCA) users (118), and nonusers (2406). After 5 years, SSRI users had the greatest bone loss (0.8% decrease in BMD). Rauma et al. reported reduced BMD in 928 men receiving SSRIs or SNRIs, but not in TCAs users [102, 104, 105].

9.1 Pathophysiology of serotonin reuptake inhibitors and bone health

Besides hypothalamic–pituitary–adrenal (HPA) axis dysregulation and other hormonal abnormalities, MDD-related lifestyle factors such as poor diet, lack of physical activity, and smoking may also contribute to bone mineralization problems [103, 106–109]. Psychotropic medications, especially selective serotonin reuptake inhibitors (SSRI), can also increase fracture risks [103–106]. Serotonin receptors, neurotransmitters, and transporters have been discovered in osteoblasts and osteoclasts since 2001. Gut bacteria synthesize 95% of serotonin [97, 110, 111]. Gut and brain serotonin was found to have different actions on the bone metabolism by acting through different pathways as follows [112]:

1. Gut-derived serotonin reduces osteoblast proliferation, which causes bone loss. In addition to providing signals to osteoblasts through its binding to the receptor Htr1b located on their surface, serotonin inhibits phosphorylation of cAMP responsive element binding protein (CREB) by phosphokinase A (PKA), resulting in decreased expression of cyclin genes and reduced osteoblast proliferation. A crucial role is played by Wnt-catenin signaling in this system since it regulates osteoblast differentiation, proliferation, survival, and bone formation [112–115].

2. Brain-derived serotonin reduces sympathetic output, which favors bone growth. On the other hand, brain-derived serotonin communicates with the ventromedial hypothalamic neurons via Htr2c receptors to decrease the sympathetic output and increase bone formation [112, 116].

3. SSRIs may act independently on osteoclast's Ca calmodulin-dependent activation of c-FoseNfatc1 cascade leading to decreased bone resorption. For the shorter duration of use of SSRI, the independent effect on the bone, i.e., decreases in bone resorption predominate, while on long-term use, both independent and serotonin-mediated effects counteract each other leading to bone loss [112, 117].

10. Diabetes and bone fragility

Both osteoporosis and diabetes mellitus are widespread chronic diseases that affect the elders. A considerable number of the population who are at risk for osteoporosis is expected to have a corresponding diabetes because the incidence of both diseases might be as high as 35–40% [118–121].

With a fracture relative risk (RR) ranging from 1.5 to 3, it is clear that type 2 diabetes is linked to a higher risk of fractures. This appears to be more prevalent in older persons with poorly controlled diabetes and a longer disease duration (>5 years) [120–124].

Diabetic osteoporosis and diabetic bone disease have been proposed, but they are still not commonly used, in persons who have had diabetes for longer than 10 years [125, 126].

10.1 The pathophysiological mechanisms underlying bone fragility in T2DM

The main causes of T2DM-induced bone fragility comprise chronic hyperglycemia, advanced glycation end (AGE) product accumulation, insulin resistance, altered bone marrow adiposity, inflammatory agents, adipokines generated by visceral fat, and oxidative stress [119].

10.1.1 Hyperglycemia

Exerts both direct and indirect effects on the osteoblastic differentiation and function. Differentiation of bone marrow mesenchymal cells is shifted toward adipogenesis rather than osteogenesis in the presence of hyperglycemia due to:

1. Upregulation of the transcription factor peroxisome proliferator activated receptor-γ (PPAR-γ), which promotes adipogenesis,

2. Downregulation of Runx2/core-binding factor α1 (Cbfα1), which regulates osteoblast differentiation and maturation [124]. Moreover, increased cytokine levels have been shown to suppress osteoblast differentiation and accelerate osteoclastogenesis [125, 126].

10.1.2 AGEs accumulation

By crosslinking with the collagen fibers in bone, AGEs harm the bone by causing microarchitectural degeneration and increased bone fragility [127]. Studies conducted in vitro have demonstrated that AGEs also enhance osteocyte sclerostin expression, a potent inhibitor of bone formation [119, 128].

10.1.3 Insulin resistance

The development of insulin resistance may be one of those deleterious effects on the bone health. It is postulated that decreased muscle strength secondary to decreased glucose uptake by muscles can compromise skeletal loading [119]. Receptors for glucagon-like peptide-1 (GLP-1) are normally also expressed on bone marrow stromal cells and immature osteoblasts, and GLP1 has been shown to stimulate the proliferation of mesenchymal stem cells and inhibit their differentiation into adipocytes [119, 129].

10.1.4 Microarchitecture abnormalities

Diabetes alone degrades the organic composition and strength of bone, which has an impact on its biomechanical qualities [124]. Long-term diabetes compromises bone collagen microstructure, mineralization, and bone strength. Increased porosity and decreased cortical density are two changes in the bone structure associated with T2DM. These elements lead to abnormal bone architecture, which lowers bone's ability to withstand mechanical stress and increases the risk of fragility fractures [119].

10.1.5 Anti-diabetes medications

Thiazolidinediones (TZDs, glitazones) enhance insulin sensitivity and beta cell response to a glucose load by acting as agonists of nuclear peroxisome proliferator-activated receptor gamma (PPAR-Y). The differentiation of precursor cells into osteoblasts depends heavily on PPAR-Y, which is expressed in osteoblasts, osteoclasts, and stromal cells of the bone marrow. Adipogenesis is enhanced by PPAR-Y activation, which also inhibits osteoblastogenesis [130]. The ADOPT trial was the first to report a link between TZDs and an elevated risk of fracture [130, 131]. When compared to metformin and glyburide treatment arms, the incidence of fracture in the lower and upper limbs was roughly twice as high in women using rosiglitazone [132]. The RECORD study displayed that rosiglitazone was associated with an increased risk of fractures (10.7%) as compared to the combination of metformin/sulfonylurea (6.8%) [130]. Both rosiglitazone and pioglitazone were associated with a significantly increased risk of fractures [133]. In men with type 2 diabetes, rosiglitazone has been linked to a higher risk of vertebral fractures [134, 135]. In the ACCORD bone research, female patients who stopped taking TZD had a reduced incidence of fracture after 1–2 years [136].

11. Conclusion

A number of medications we use on a daily basis can negatively affect bone health and lead to bone loss. Collagen and other organic compounds of the matrix may also be affected, as well as mineral density, trabecular structure, and hydroxyapatite. Additionally, cellular turnover may be affected, leading to an imbalance between bone formation and resorption as well as between organic and inorganic matrix composition. Consequently, the biomechanical ability of bone is diminished, resulting in decreased bone density and an increased risk of fracture. Additionally, even in the presence of normal bone biomechanical competence, some drugs may increase the risk of falls and fractures.

A variety of drugs may cause bone loss by: 1. Lowering estrogen levels (e.g., aromatase inhibitors used in breast cancer, GnRH agonists used in prostate cancer). 2. Interfering with vitamin D levels (Enzyme-Inducing Anti-Seizure Medications), EIASM therapy has been proven to cause increased vitamin D metabolism, low vitamin D levels, impaired calcium absorption, and resultant hypocalcemia leading to secondary hyperparathyroidism and increased osteoclastic bone resorption. 3. Directly affecting bone cells (chemotherapy, phenytoin, or thiazolidinediones) which divert mesenchymal stem cells from osteoblastogenesis to adipocytogenesis, consequently, an imbalance occurs between bone formation and resorption, as well as between soft organic matrix and hard inorganic matrix. Besides effects on the mineralized matrix, interactions with collagen and other nonmineralized matrix components can decrease bone biomechanical competence without affecting bone mineral density (BMD). 4. Thyroid dysfunction caused by long-term use of (anti-arrhythmic drug amiodarone) that can affect bone health might happen, though as follows: 1- Amiodarone-induced thyrotoxicosis (AIT) or 2- Amiodarone-induced hypothyroidism (AIH). 5. Strong inhibitors of stomach acid secretion (PPIs), which is thought to be important for calcium absorption by enhancing the solubility of calcium salts that are insoluble leading to decreases intestinal calcium absorption, and ultimately causing a decline in bone mineral density. 6. MTX treatment may reduce osteoblastic differentiation, which would then lead to a reduction in bone formation and an increased risk of osteopathy. To clarify the long-term effects of MTX on bone density of both axial and peripheral bones, a longitudinal investigation is consequently required. 7. Patients with IBD are more likely than general population to experience bone loss due to malnutrition, vitamin D and calcium malabsorption and deficiency, vitamin K insufficiency, immobilization, underlying inflammatory state. 8. High blood pressure is linked to increased urinary calcium loss, which impairs the calcium balance necessary for bone remodeling. Furthermore, high levels of the parathyroid hormone are linked to hypertension and accelerated bone turnover, reducing bone mass and bone quality. Finally, high blood pressure may gradually harm brain regions involved in balance and gait regulation, which could increase the risk of falls and consequent fractures. 9. Psychotropic medications, especially selective serotonin reuptake inhibitors (SSRIs), can increase fracture risk. Gut-derived serotonin reduces osteoblast proliferation, which causes bone loss. 10. Thiazolidinediones (TZDs, Glitazones) enhance insulin sensitivity and beta cell response to a glucose load by acting as agonists of nuclear peroxisome proliferator-activated receptor gamma (PPAR-Y), leading to enhanced adipogenesis and inhibited osteoblastogenesis.

Author details

Hesham Hamoud
Al Azhar University, Cairo, Egypt

*Address all correspondence to: hamoud.hesham@yahoo.com

References

[1] Tsadok MA, Jackevicius CA, Rahme E, Essebag V, Eisenberg MJ, Humphries KH, et al. Amiodarone-induced thyroid dy sfunction: Brand-name versus generic formulations. Canadian Medical Association Journal. 2011;**183**(12):E817-E823. DOI: 10.1503/cmaj.101800 [Medline]. [Full Text] PMID 21746822

[2] Diederichsen SZ, Darkner S, Chen X, Johannesen A, Pehrson S, Hansen J, et al. Short-term amiodarone treatment for atrial fibrillation after catheter ablation induces a transient thyroid dysfunction: Results from the placebo-controlled, randomized AMIO-CAT trial. European Journal of Internal Medicine. 2016;**33**:36-41. DOI: 10.1016/j.ejim.2016.04.012 [Medline] PMID 27129624

[3] Ahmed S, Van Gelder IC, Wiesfeld AC, Van Veldhuisen DJ, Links TP. Determinants and outcome of amiodarone-associated thyroid dysfunction. Clinical Endocrinology. 2011;**75**(3):388-394. [Medline], PMID 21535072. DOI: 10.1111/j.1365-2265.2011.04087.x

[4] Mosher MC. Amiodarone-induced hypothyroidism and other adverse effects. Dimensions of Critical Care Nursing. 2011;**30**(2):87-93. [Medline] PMID 21307683. DOI: 10.1097/DCC.0b013e3182052130

[5] Hofmann A, Nawara C, Ofluoglu S, et al. Incidence and predictability of amiodarone-induced thyrotoxicosis and hypothyroidism. Wiener Klinische Wochenschrift. 2008;**120**(15-16):493-498. [Medline]. PMID 18820854. DOI: 10.1007/s00508-008-1017-2

[6] Wang M, Cai S, Sun L, Zhao Q, Feng W. Safety and efficacy of early radiofrequency catheter ablation in patients with paroxysmal atrial fibrillation complicated with amiodarone-induced thyrotoxicosis. Cardiology Journal. 2016;**23**(4):416-421. DOI: 10.5603/CJ.a2016.0029 [Medline]. PMID 27296156

[7] Kinoshita S, Hayashi T, Wada K, et al. Risk factors for amiodarone-induced thyroid dysfunction in Japan. Journal of Arrhythmia. 2016;**32**(6):474-480. DOI: 10.1016/j.joa.2016.03.008 [Medline]. [Full Text] PMID 27920832

[8] Zhong B, Wang Y, Zhang G, Wang Z. Environmental iodine content, female sex and age are associated with new-onset amiodarone-induced hypothyroidism: A systematic review and meta-analysis of adverse reactions of amiodarone on the thyroid. Cardiology. 2016;**134**(3):366-371. DOI: 10.1159/000444578 [Medline]. PMID 27100205

[9] Piga M, Serra A, Boi F, Tanda ML, Martino E, Mariotti S. Amiodarone-induced thyrotoxicosis. A review. Minerva Endocrinologica. 2008;**33**(3):213-228 [Medline]. PMID 18846027

[10] Piga M, Cocco MC, Serra A, Boi F, Loy M, Mariotti S. The usefulness of 99mTc-sestaMIBI thyroid scan in the differential diagnosis and management of amiodarone-induced thyrotoxicosis. European Journal of Endocrinology. 2008;**159**(4):423-429. DOI: 10.1530/EJE-08-0348 [Medline]. PMID 18603573

[11] Kushner PR, Peura DA. Review of proton pump inhibitors for the initial treatment of heartburn: Is there a dose ceiling effect? Advances in Therapy. 2011;**28**(5):367-388. DOI: 10.1007/s12325-011-0013-5 PMID 21484335

[12] Mönnikes H, Schwan T, van Rensburg C, Straszak A, Theek C,

Lühmann R, et al. Possible etiology of improvements in both quality of life and overlapping gastroesophageal reflux disease by proton pump inhibitor treatment in a prospective randomized controlled trial. BMC Gastroenterology. 2013;**13**:145. DOI: 10.1186/1471-230X-13-145 PMID 24083350

[13] Tsoi KK, Hirai HW, Sung JJ. Meta-analysis: Comparison of oral vs. intravenous proton pump inhibitors in patients with peptic ulcer bleeding. Alimentary Pharmacology & Therapeutics. 2013;**38**(7):721-728. DOI: 10.1111/apt.12441 PMID 23915096

[14] Ye X, Liu H, Wu C, Qin Y, Zang J, Gao Q, et al. Proton pump inhibitors therapy and risk of hip fracture: A systematic review and meta-analysis. European Journal of Gastroenterology & Hepatology. 2011;**23**(9):794-800. DOI: 10.1097/MEG.0b013e328348a56a PMID 21701389

[15] Yu EW, Bauer SR, Bain PA, Bauer DC. Proton pump inhibitors and risk of fractures: A meta-analysis of 11 international studies. The American Journal of Medicine. 2011;**124**(6):519-526. DOI: 10.1016/j.amjmed.2011.01.007 PMID 21605729

[16] Eom CS, Park SM, Myung SK, Yun JM, Ahn JS. Use of acid-suppressive drugs and risk of fracture: A meta-analysis of observational studies. Annals of Family Medicine. 2011;**9**(3):257-267. DOI: 10.1370/afm.1243 PMID 21555754

[17] Ngamruengphong S, Leontiadis GI, Radhi S, Dentino A, Nugent K. Proton pump inhibitors and risk of fracture: A systematic review and meta-analysis of observational studies. The American Journal of Gastroenterology. 2011;**106**(7):1209-1218; quiz 1219 [quiz]. DOI: 10.1038/ajg.2011.113 PMID 21483462

[18] FDA Drug Safety Communication. Possible Increased Risk of Fractures of the Hip, Wrist, and Spine with the Use of Proton Pump Inhibitors. Rockville: U.S. Food and Drug Administration; 2010

[19] Famularo G, Gasbarrone L, Minisola G. Hypomagnesemia and proton-pump inhibitors. Expert Opinion on Drug Safety. 2013;**12**(5):709-716. DOI: 10.1517/14740338.2013.809062 PMID 23808631

[20] Zhou B, Huang Y, Li H, Sun W, Liu J. Proton-pump inhibitors and risk of fractures: An update meta-analysis. Osteoporosis International. 2016;**27**(1):339-347. DOI: 10.1007/s00198-015-3365-x PMID 26462494

[21] Lintermans A, Neven P. Safety of aromatase inhibitor therapy in breast cancer. Expert Opinion on Drug Safety. 2015;**14**(8):1201-1211. DOI: 10.1517/14740338.2015.1053458 PMID 26059833

[22] Chien AJ, Goss PE. Aromatase inhibitors and bone health in women with breast cancer. Journal of Clinical Oncology. 2006;**24**(33):5305-5312. DOI: 10.1200/JCO.2006.07.5382 PMID 17114665

[23] Suskin J, Shapiro CL. Osteoporosis and musculoskeletal complications related to therapy of breast cancer. Gland Surgery. 2018;**7**(4):411-423. DOI: 10.21037/gs.2018.07.05 PMID 30175057

[24] Napoli N, Rastelli A, Ma C, Yarramaneni J, Vattikutti S, Moskowitz G, et al. Genetic polymorphism at Val80 (rs700518) of the CYP19A1 gene is associated with aromatase inhibitor associated bone loss in women with ER breast cancer. Bone. 2013;**55**(2):309-314. DOI: 10.1016/j.bone.2013.04.021 PMID 23643682

[25] Oesterreich S, Henry NL, Kidwell KM, Van Poznak CH, Skaar TC, Dantzer J, et al. Associations between genetic variants and the effect of

letrozole and exemestane on bone mass and bone turnover. Breast Cancer Research and Treatment. 2015;**154**(2):263-273. DOI: 10.1007/s10549-015-3608-8 PMID 26536870

[26] Rodríguez-Sanz M, García-Giralt N, Prieto-Alhambra D, Servitja S, Balcells S, Pecorelli R, et al. CYP11A1 expression in bone is associated with aromatase inhibitor-related bone loss. Journal of Molecular Endocrinology. 2015;**55**(1):69-79. DOI: 10.1530/JME-15-0079 PMID 26108486

[27] Liu M, Goss PE, Ingle JN, Kubo M, Furukawa Y, Batzler A, et al. Aromatase inhibitor-associated bone fractures: A case-cohort GWAS and functional genomics. Molecular Endocrinology. 2014;**28**(10):1740-1751. DOI: 10.1210/me.2014-1147 PMID 25148458

[28] Wang J, Lu K, Song Y, Zhao S, Ma W, Xuan Q et al. RANKL and OPG polymorphisms are associated with aromatase inhibitor-related musculoskeletal adverse events in Chinese Han breast cancer patients. PLoS One. 2015;**10**(7):e0133964. DOI: 10.1371/journal.pone.0133964, PMID 26218592

[29] May KP, West SG, Mcdermott MT, Huffer WE. The effect of low dose methotrexate on bone metabolism and histomorphometry in rats. Arthritis and Rheumatism. 1994;**37**(2):201-206. DOI: 10.1002/art.1780370208 PMID 8129775

[30] Preston SJ, Diamond T, Scott A, Laurent MR. Methotrexate osteopathy in rheumatic disease. Annals of the Rheumatic Diseases. 1993;**52**(8):582-585. DOI: 10.1136/ard.52.8.582 PMID 8215620

[31] Wang D, Christensen K, Chawla K, Xiao G, Krebsbach PH, Franceschi RT. Isolationandcharacterizationof MC3T3-E1 preosteoblast subclones with distinct in vitro and in vivo differentiation/mineralization potential. Journal of Bone and Mineral Research. 1999;**14**(6):893-903. DOI: 10.1359/jbmr.1999.14.6.893 PMID 10352097

[32] Wang D, Christensen K, Chawla K, Xiao G, Krebsbach PH, Franceschi RT. Isolationandcharacterizationof MC3T3-E1 preosteoblast subclones with distinct in vitro and in vivo differentiation/mineralization potential. Journal of Bone and Mineral Research. 1999;**14**(6):893-903. DOI: 10.1359/jbmr.1999.14.6.893, PMID 10352097

[33] May KP, Mercill D, Mcdermott MT, West SG. The effect of methotrexate on mouse bone cells in culture. Arthritis and Rheumatism. 1996;**39**(3):489-494. DOI: 10.1002/art.1780390317 PMID 8607898

[34] Zonneveld IM, Bakker WK, Dijkstra PF, Bos JD, van Soesbergen RM, Dinant HJ. Methotrexate osteopathy in long-term, low-dose methotrexate treatment for psoriasis and rheumatoid arthritis. Archives of Dermatology. 1996;**132**(2):184-187 PMID 8629827.

[35] Ott C, Schölmerich J. Extraintestinal manifestations and complications in IBD. Nature Reviews. Gastroenterology & Hepatology. 2013;**10**(10):585-595. DOI: 10.1038/nrgastro.2013.117 PMID 23835489 [PMID 23835489]

[36] Miheller P, Gesztes W, Lakatos PL. Manipulating bone disease in inflammatory bowel disease patients. Annals of Gastroenterology. 2013;**26**(4):296-303 PMID 24714303

[37] Rothfuss KS, Stange EF, Herrlinger KR. Extraintestinal manifestations and complications in inflammatory bowel diseases. World Journal of Gastroenterology. 2006;**12**(30):4819-4831. DOI: 10.3748/wjg.v12.i30.4819 PMID 16937463

[38] Ali T, Lam D, Bronze MS, Humphrey MB. Osteoporosis in inflammatory bowel disease. The American Journal of Medicine. 2009;**122**(7):599-604. DOI: 10.1016/j.amjmed.2009.01.022 PMID 19559158 [PMID19559158]

[39] Targownik LE, Bernstein CN, Leslie WD. Inflammatory bowel disease and the risk of osteoporosis and fracture. Maturitas. 2013;**76**(4):315-319. DOI: 10.1016/j.maturitas.2013.09.009 PMID 24139749 [PMID 24139749]

[40] Miheller P, Lorinczy K, Lakatos PL. Clinical relevance of changes in bone metabolism in inflammatory bowel disease. World Journal of Gastroenterology. 2010;**16**:5536-5542 [PMID 21105186, p. i44.5536]. DOI: 10.3748/wjg

[41] Bernstein CN, Blanchard JF, Leslie W, Wajda A, Yu BN. The incidence of fracture among patients with inflammatory bowel disease. A population-based cohort study. Annals of Internal Medicine. 2000;**133**(10):795-799. DOI: 10.7326/0003-4819-133-10-200011210-00012 PMID 11085842 [PMID 11085842]

[42] Schoon EJ, Blok BM, Geerling BJ, Russel MG, Stockbrügger RW, Brummer RJ. Bone mineral density in patients with recently diagnosed inflammatory bowel disease. Gastroenterology. 2000;**119**(5):1203-1208. DOI: 10.1053/gast.2000.19280 PMID 11054377 [PMID 11054377]

[43] Sakellariou GT, Moschos J, Berberidis C, Mpoumponaris A, Kadis S, Molyvas E, et al. Bone density in young males with recently diagnosed inflammatory bowel disease. Joint, Bone, Spine. 2006;**73**(6):725-728. DOI: 10.1016/j.jbspin.2006.01.017 PMID 17126059 [PMID 17126059]

[44] Reinshagen M. Osteoporosis in inflammatory bowel disease. Journal of Crohn's & Colitis. 2008;**2**:202-207 [PMID 21172211]. DOI: 10.1016/j.crohns.2008.01.005

[45] Mundy GR. Osteoporosis and inflammation. Nutrition Reviews. 2007;**65**(12 Pt 2):S147-S151. DOI: 10.1111/j.1753-4887.2007.tb00353.x PMID 18240539 [PMID 18240539]

[46] Redlich K, Smolen JS. Inflammatory bone loss: Pathogenesis and therapeutic intervention. Nature Reviews. Drug Discovery. 2012;**11**(3):234-250. DOI: 10.1038/nrd3669 PMID 22378270 [PMID 22378270]

[47] Reffitt DM, Meenan J, Sanderson JD, Jugdaohsingh R, Powell JJ, Thompson RP. Bone density improves with disease remission in patients with inflammatory bowel disease. European Journal of Gastroenterology & Hepatology. 2003;**15**(12):1267-1273. DOI: 10.1097/00042737-200312000-00003 PMID 14624148 [PMID 14624148]

[48] Wong PK, Christie JJ, Wark JD. The effects of smoking on bone health. Clinical Science (London, England). 2007;**113**(5):233-241. DOI: 10.1042/CS20060173 PMID 17663660 [PMID 17663660]

[49] Yoon V, Maalouf NM, Sakhaee K. The effects of smoking on bone metabolism. Osteoporosis International. 2012;**23**(8):2081-2092. DOI: 10.1007/s00198-012-1940-y PMID 22349964 [PMID 22349964]

[50] Cosnes J. Smoking, physical activity, nutrition and lifestyle: Environmental factors and their impact on IBD. Digestive Diseases. 2010;**28**(3):411-417. DOI: 10.1159/000320395 PMID 20926865 [PMID 20926865]

[51] Silvennoinen J, Lamberg-Allardt C, Kärkkäinen M, Niemelä S, Lehtola J.

Dietary calcium intake and its relation to bone mineral density in patients with inflammatory bowel disease. Journal of Internal Medicine. 1996;**240**(5):285-292. DOI: 10.1046/j.1365-2796.1996.25862000.x PMID 8946811

[52] Cashman KD, Shanahan F. Is nutrition an aetiological factor for inflammatory bowel disease? European Journal of Gastroenterology & Hepatology. 2003;**15**(6):607-613. DOI: 10.1097/00042737-200306000-00005 PMID 12840670

[53] McCarthy D, Duggan P, O'Brien M, Kiely M, McCarthy J, Shanahan F, et al. Seasonality of vitamin D status and bone turnover in patients with Crohn's disease. Alimentary Pharmacology & Therapeutics. 2005;**21**(9):1073-1083. DOI: 10.1111/j.1365-2036.2005.02446.x PMID 15854168

[54] Gilman J, Shanahan F, Cashman KD. Determinants of vitamin D status in adult Crohn's disease patients, with particular emphasis on supplemental vitamin D use. European Journal of Clinical Nutrition. 2006;**60**(7):889-896. DOI: 10.1038/sj.ejcn.1602395 PMID 16493452

[55] Mouli VP, Ananthakrishnan AN. Review article: Vitamin D and inflammatory bowel diseases. Alimentary Pharmacology & Therapeutics. 2014;**39**(2):125-136. DOI: 10.1111/apt.12553 PMID 24236989 [PMID 24236989]

[56] Duggan P, O'Brien M, Kiely M, McCarthy J, Shanahan F, Cashman KD. Vitamin K status in patients with Crohn's disease and relationship to bone turnover. The American Journal of Gastroenterology. 2004;**99**(11):2178-2185. DOI: 10.1111/j.1572-0241.2004.40071.x PMID 15555000

[57] Dussault PM, Lazzari AA. Epilepsy and osteoporosis risk. Current Opinion in Endocrinology, Diabetes, and Obesity. 2017;**24**(6):395-401. DOI: 10.1097/MED.0000000000000366 PMID 28857845

[58] Shiek Ahmad B, Hill KD, O'Brien TJ, Gorelik A, Habib N, Wark JD. Falls and fractures in patients chronically treated with antiepileptic drugs. Neurology. 2012;**79**(2):145-151. DOI: 10.1212/WNL.0b013e31825f0466 PMID 22700806

[59] Samaniego EA, Sheth RD. Bone consequences of epilepsy and antiepileptic medications. Seminars in Pediatric Neurology. 2007;**14**(4):196-200. DOI: 10.1016/j.spen.2007.08.006 PMID 18070676

[60] Beerhorst K, Tan IY, De Krom M, Verschuure P, Aldenkamp AP. Antiepileptic drugs and high prevalence of low bone mineral density in a group of inpatients with chronic epilepsy. Acta Neurologica Scandinavica. 2013;**128**(4):273-280. DOI: 10.1111/ane.12118 PMID 23461582

[61] Vestergaard P, Tigaran S, Rejnmark L, Tigaran C, Dam M, Mosekilde L. Fracture risk is increased in epilepsy. Acta Neurologica Scandinavica. 1999;**99**(5):269-275. DOI: 10.1111/j.1600-0404.1999.tb00675.x PMID 10348155

[62] Vestergaard P. Epilepsy, osteoporosis and fracture risk - a meta-analysis. Acta Neurologica Scandinavica. 2005;**112**(5):277-286. DOI: 10.1111/j.1600-0404.2005.00474.x PMID 16218908

[63] Desai KB, Ribbans WJ, Taylor GJ. Incidence of five common fracture types in an institutional epileptic population. Injury. 1996;**27**(2):97-100. DOI: 10.1016/0020-1383(95)00189-1 PMID 8730381

[64] Jetté N, Lix LM, Metge CJ, Prior HJ, McChesney J, Leslie WD. Association of

antiepileptic drugs with nontraumatic fractures: A population-based analysis. Archives of Neurology. 2011;**68**(1):107-112. DOI: 10.1001/archneurol.2010.341 PMID 21220681

[65] Miziak B, Chrościńska-Krawczyk M, Czuczwar SJ. An update on the problem of osteoporosis in people with epilepsy taking antiepileptic drugs. Expert Opinion on Drug Safety. 2019;**18**(8):679-689. DOI: 10.1080/14740338.2019.1625887 PMID 31159612

[66] Dent CE, Richens A, Rowe DJ, Stamp TC. Osteomalacia with long-term anticonvulsant therapy in epilepsy. British Medical Journal. 1970;**4**(5727):69-72. DOI: 10.1136/bmj.4.5727.69 PMID 5312062

[67] Svalheim S, Sveberg L, Mochol M, Taubøll E. Interactions between antiepileptic drugs and hormones. Seizure. 2015;**28**:12-17. DOI: 10.1016/j.seizure.2015.02.022 PMID 25797888

[68] Krishnamoorthy G, Nair R, Sundar U, Kini P, Shrivastava M. Early predisposition to osteomalacia in Indian adults on phenytoin or valproate monotherapy and effective prophylaxis by simultaneous supplementation with calcium and 25-hydroxy vitamin D at recommended daily allowance dosage: A prospective study. Neurology India. 2010;**58**(2):213-219. DOI: 10.4103/0028-3886.63796 PMID 20508338

[69] Hahn TJ, Halstead LR. Anticonvulsant drug-induced osteomalacia: Alterations in mineral metabolism and response tovitamin D3 administration. Calcified Tissue International. 1979;**27**(1):13-18. DOI: 10.1007/BF02441155 PMID 223750

[70] Tjellesen L, Nilas L, Christiansen C. Does carbamazepine cause disturbances in calcium metabolism in epileptic patients? Acta Neurologica Scandinavica. 1983;**68**(1):13-19. DOI: 10.1111/j.1600-0404.1983.tb04809.x PMID 6613523

[71] Kwasa JK, Amayo A, Ndavi PM, Kwasa TO. Bone metabolism in healthy ambulatory control premenopausal women and in epileptics on anti-convulsant drugs. East African Medical Journal. 2010;**87**(4):151-155. DOI: 10.4314/eamj.v87i4.62411 PMID 23057290

[72] Ashjazadeh N, Zamani A, Pourjafar M, Omrani GR. Bone density of ambulatory adult patients receiving long-term anticonvulsant drug therapy. Archives of Iranian Medicine. 2009;**12**(6):550-554 PMID 19877746

[73] Hajjar I, Kotchen TA. Trends in prevalence, awareness, treatment, and control of hypertension in the United States, 1988-2000. Journal of the American Medical Association. 2003;**290**(2):199-206. DOI: 10.1001/jama.290.2.199 PMID 12851274

[74] Kanis JA, on behalf of the World Health Organization Scientific Group (2007). Assessment of osteoporosis at the primary health-care level. In: Technical Report. World Health Organization Collaborating Centre for Metabolic Bone Diseases. UK: University of Sheffield; 2007

[75] Yang S, Nguyen ND, Center JR, Eisman JA, Nguyen TV. Association between hypertension and fragility fracture: A longitudinal study. Osteoporosis International. 2014;**25**(1):97-103. DOI: 10.1007/s00198-013-2457-8 PMID 23892585

[76] Cummings SR, Nevitt MC, Browner WS, Stone K, Fox KM, Ensrud KE, et al. Risk factors for hip fracture in white women. Study of osteoporotic fractures research group. The New England Journal of Medicine. 1995;**332**(12):767-773. DOI: 10.1056/NEJM199503233321202 PMID 7862179

[77] Kannel WB. Risk factors in hypertension. Journal of Cardiovascular Pharmacology. 1989;**13**(Suppl 1):S4-S10. DOI: 10.1097/00005344-198900131-00003 PMID 2468976

[78] MacGregor GA, Cappuccio FP. The kidney and essential hypertension: A link to osteoporosis? Journal of Hypertension. 1993;**11**(8):781-785. DOI: 10.1097/00004872-199308000-00003 PMID 8228200

[79] Gadallah M, Massry SG, Bigazzi R, Horst RL, Eggena P, Campese VM. Intestinal absorption of calcium and calcium metabolism in patients with essential hypertension and normal renal function. American Journal of Hypertension. 1991;**4**(5 Pt 1):404-409. DOI: 10.1093/ajh/4.5.404, PMID 2069773

[80] Young EW, Morris CD, McCarron DA. Urinary calcium excretion in essential hypertension. Journal of Laboratory and Clinical Medicine. 1992;**120**(4):624-632 PMID 1402338

[81] Cappuccio FP, Meilahn E, Zmuda JM, Cauley JA. High blood pressure and bone-mineral loss in elderly whitewomen: a prospective study. Lancet. 1999;**354**(9183):971-975. DOI: 10.1016/s0140-6736(99)01437-3 PMID 10501357

[82] Tsuda K, Nishio I, Masuyama Y. Bone mineral density in women with essential hypertension. American Journal of Hypertension. 2001;**14**(704-707):704-707. DOI: 10.1016/s0895-7061(01)01303-6 PMID 11465657

[83] Vestergaard P, Rejnmark L, Mosekilde L. Hypertension is a risk factor for fractures. Calcified Tissue International. 2009;**84**(2):103-111. DOI: 10.1007/s00223-008-9198-2 PMID 19067019

[84] Sennerby U, Melhus H, Gedeborg R, Byberg L, Garmo H, Ahlbom A, et al. Cardiovascular diseases and risk of hip fracture. Journal of the American Medical Association. 2009;**302**(15):1666-1673. DOI: 10.1001/jama.2009.1463 PMID 19843901

[85] Pérez-Castrillón JL, Martín-Escudero JC, Alvarez Manzanares P, Cortés Sancho R, Iglesias Zamora S, García AM. Hypertension as a risk factor for hip fracture. American Journal of Hypertension. 2005;**18**(1):146-147. DOI: 10.1016/j.amjhyper.2004.08.016 PMID 15691630

[86] McCarron DA, Pingree PA, Rubin RJ, Gaucher SM, Molitch M, Krutzik S. Enhanced parathyroid function in essential hypertension:A homeostatic response to a urinary calcium leak. Hypertension. 1980;**2**(2):162-168. DOI: 10.1161/01.hyp.2.2.162 PMID 7380520

[87] Strazzullo P, Nunziata V, Cirillo M, Giannattasio R, Ferrara LA, Mattioli PL, et al. Abnormalities of calcium metabolism in essential hypertension. Clinical Science (London, England). 1983;**65**(2):137-141. DOI: 10.1042/cs0650137 PMID 6305548

[88] Cappuccio FP, Kalaitzidis R, Duneclift S, Eastwood JB. Unravelling the links between calcium excretion, salt intake, hypertension, kidney stones and bone metabolism. Journal of Nephrology. 2000;**13**(3):169-177 PMID 10928292

[89] Rosano C, Longstreth WT Jr, Boudreau R, Taylor CA, Du Y, Kuller LH, et al. High blood pressure accelerates gait slowing in well-functioning older adults over 18-years of follow-up. Journal of the American Geriatrics Society. 2011;**59**(3):390-397. DOI: 10.1111/j.1532-5415.2010.03282.x PMID 21391929

[90] Bergland A, Jarnlo GB, Laake K. Predictors of falls in the elderly by location. Aging Clinical and

Experimental Research. 2003;**15**(1):43-50. DOI: 10.1007/BF03324479 PMID 12841418

[91] Schweiger JU, Schweiger U, Hüppe M, Kahl KG, Greggersen W, Fassbinder E. Bone density and depressive disorder: A meta-analysis. Brain and Behavior: A Cognitive Neuroscience Perspective. 2016;**6**(8):e00489. DOI: 10.1002/brb3.489 PMID 27547495

[92] Yirmiya R, Bab I. Major depression is a risk factor for low bone mineral density: A meta-analysis. Biological Psychiatry. 2009;**66**(5):423-432

[93] Klibanski A, Adams-Campbelll L, Bassford T, Blair SN, Boden SD, Dickersin K, et al. NIH Consensus Development Panel on Osteoporosis Prevention, Diagnosis, and Therapy. Osteoporosis prevention, diagnosis, and therapy. Journal of the American Medical Association. 2001;**285**(6):785-795. DOI: 10.1001/jama.285.6.785 PMID 11176917. Available from: http://consensus.nih.gov

[94] Weaver CM, Gordon CM, Janz KF, Kalkwarf HJ, Lappe JM, Lewis R, et al. Osteoporosis Foundation's position statement on peak bone mass development and lifestyle factors:A systematic review and implementation recommendations. Osteoporosis International. 2016;**27**(4):1281-1386. DOI: 10.1007/s00198-015-3440-3 PMID 26856587

[95] Cizza G, Primma S, Csako G. Depression as a risk factor for osteoporosis. Trends in Endocrinology and Metabolism. 2009, 2009;**20**(8):367-373. DOI: 10.1016/j.tem.2009.05.003 PMID 19747841

[96] Haney EM, Warden SJ, Bliziotes MM. Effects of selective serotonin reuptake inhibitors on bone health in adults: Time for recommendations about screening, prevention and management? Bone .Research support, N.I.H., extramural review. 2010;**46**(1):13-17

[97] Ortuño MJ, Robinson ST, Subramanyam P, Paone R, Huang YY, Guo XE, et al. Serotonin-reuptake inhibitors act centrally to cause bone loss in mice by counteracting a local anti-resorptive effect. Nature Medicine. 2016;**22**(10):1170e9. DOI: 10.1038/nm.4166 PMID 27595322

[98] Rawson KS, Dixon D, Civitelli R, Peterson TR, Mulsant BH, Reynolds CF 3rd, et al. Bone turnover with venlafaxine treatment in older adults with depression. Journal of the American Geriatrics Society. 2017;**65**(9):2057-2063. DOI: 10.1111/jgs.14936 PMID 28555718

[99] Warden SJ, Robling AG, Haney EM, Turner CH, Bliziotes MM. The emerging role of serotonin (5-hydroxytryptamine) in the skeleton and its mediation of the skeletal effects of low-density lipoprotein receptor-related protein 5 (LRP5). Bone. 2010;**46**(1):4-12. DOI: 10.1016/j.bone.2009.06.029 PMID 19591966

[100] Leipzig RM, Cumming RG, Tinetti ME. Drugs and falls in older people: A sys thematic review and meta-analysis: I. Psychotropic drugs. Journal of the American Geriatrics Society. 1999;**47**:30.e9

[101] Rabenda V, Nicolet D, Beaudart C, Bruyère O, Reginster JY. Relationship between use of antidepressants and risk of fractures: A meta-analysis. Osteoporosis International. 2013;**24**(1):121-137. DOI: 10.1007/s00198-012-2015-9 PMID 22638709

[102] Rauma PH, Pasco JA, Berk M, Stuart AL, Koivumaa-Honkanen H, Honkanen RJ, et al. The association between major depressive disorder, use of antidepressants and bone

mineral density (BMD) in men. Journal of Musculoskeletal and Neuronal Interactions. 2015;**15**:177.e85

[103] Hung SC, Lin CH, Hung HC, Lin CL, Lai SW. Use of selective serotonin reuptake inhibitors and risk of hip fracture in the elderly: A case-control study in Taiwan. Journal of the American Medical Directors Association. 2017;**18**(4):350-354. DOI: 10.1016/j.jamda.2016.12.003 PMID 28159466

[104] Liu B, Anderson G, Mittmann N, To T, Axcell T, Shear N. Use of selective serotonin-reuptake inhibitors or tricyclic antidepressants and risk of hip fractures in elderly people. Lancet. 1998;**351**(9112):1303-1307. DOI: 10.1016/s0140-6736(97)09528-7 PMID 9643791

[105] Vestergaard P, Rejnmark L, Mosekilde L. Anxiolytics, sedatives, antidepressants, neuroleptics and the risk of fracture. Osteoporosis International. 2006;**17**(6):807-816. DOI: 10.1007/s00198-005-0065-y PMID 16520889

[106] Vestergaard P, Rejnmark L, Mosekilde L. Selective serotonin reuptake inhibitors and other antidepressants and risk of fracture. Calcified Tissue International. 2008;**82**(2):92-101. DOI: 10.1007/s00223-007-9099-9 PMID 18219438

[107] Gagne JJ, Patrick AR, Mogun H, Solomon DH. Antidepressants and fracture risk in older adults: A comparative safety analysis. Clinical Pharmacology and Therapeutics. 2011;**89**(6):880-887. DOI: 10.1038/clpt.2011.54 PMID 21508938

[108] van den Brand MW, Pouwels S, Samson MM, van Staa TP, Thio B, Cooper C, et al. Use of antidepressants and the risk of fracture of the hip or femur. Osteoporosis International. 2009;**20**:1705e13

[109] Moura C, Bernatsky S, Abrahamowicz M, Papaioannou A, Bessette L, Adachi J, et al. Antidepressant use and 10-year incident fracture risk: The population based Canadian multicentre osteoporosis study (CaMoS). Osteoporosis International. 2014;**25**(5):1473-1481. DOI: 10.1007/s00198-014-2649-x PMID 24566587

[110] Haney EM, Chan BK, Diem SJ, Ensrud KE, Cauley JA, Barrett-Connor E, et al. Association of low bone mineral density with selective serotonin reuptake inhibitor use by older men. Archives of Internal Medicine. 2007;**167**(12):1246-1251. DOI: 10.1001/archinte.167.12.1246 PMID 17592097

[111] Crowell MD. Role of serotonin in the pathophysiology of the irritable bowel syndrome. British Journal of Pharmacology. 2004;**141**(8):1285-1293. DOI: 10.1038/sj.bjp.0705762 PMID 15100164

[112] Abid S, Houssaini A, Chevarin C, Marcos E, Tissot CM, Gary-Bobo G, et al. Inhibition of gut- and lung-derived serotonin attenuates pulmonary hypertension in mice. American Journal of Physiology. Lung Cellular and Molecular Physiology. 2012;**303**(6):L500-L508. DOI: 10.1152/ajplung.00049.2012 PMID 22797248

[113] Jedlitschky G, Greinacher A, Kroemer HK. Transporters in human platelets: Physiologic function and impact for pharmacotherapy. Blood. 2012;**119**(15):3394-3402. DOI: 10.1182/blood-2011-09-336933 PMID 22337717

[114] Ducy P, Karsenty G. The two faces of serotonin in bone biology. The Journal of Cell Biology. 2010;**191**(1):7.e13. DOI: 10.1083/jcb.201006123 PMID 20921133

[115] Oury F, Yadav VK, Wang Y, Zhou B, Liu XS, Guo XE, et al. CREB mediates

brain serotonin regulation of bone mass through its expression in ventromedial hypothalamic neurons. Genes Dev. 2010;**24**(20):2330-2342. DOI: 10.1101/gad.1977210 PMID 20952540

[116] Li F, Chong ZZ, Maiese K. Vital elements of the Wnt-frizzled signaling pathway in the nervous system. Current Neurovascular Research. 2005;**2**(4):331-340. DOI: 10.2174/156720205774322557 PMID 16181124

[117] Yadav VK, Ryu JH, Suda N, Tanaka KF, Gingrich JA, Schütz G, et al. Lrp5 controls bone formation by inhibiting serotonin synthesis in the duodenum. Cell. 2008;**135**(5):825-837. DOI: 10.1016/j.cell.2008.09.059 PMID 19041748

[118] Deepa M, Grace M, Binukumar B, Pradeepa R, Roopa S, Khan HM, et al. High burden of prediabetes and diabetes in three large cities in South Asia: The center for cardio-metabolic risk reduction in South Asia (CARRS) study. Diabetes Research and Clinical Practice. 2015;**110**(2):172.e82. DOI: 10.1016/j.diabres.2015.09.005 PMID 26432412

[119] Napoli N, Chandran M, Pierroz DD, Abrahamsen B, Schwartz AV, Ferrari SL, et al. Mechanisms of diabetes mellitus-induced bone fragility. Nature Reviews. Endocrinology. 2017;**13**(4):208-219. DOI: 10.1038/nrendo.2016.153 PMID 27658727

[120] Ferrari SL, Abrahamsen B, Napoli N, Akesson K, Chandran M, Eastell R, et al. Diagnosis and management of bone fragility in diabetes: An emerging challenge. Osteoporosis International. 2018;**29**(12):2585-2596. DOI: 10.1007/s00198-018-4650-2 PMID 30066131

[121] Wang H, Ba Y, Xing Q, Du JL. Diabetes mellitus and the risk of fractures at specific sites: A meta-analysis. BMJ Open. 2019;**9**(1):e024067. DOI: 10.1136/bmjopen-2018-024067 PMID 30610024

[122] Bouillon R. Diabetic bone disease. Calcified Tissue International. 1991;**49**:155.e60. DOI: 10.1007/BF02556109

[123] Murray CE, Coleman CM. Impact of diabetes mellitus on bone health. International Journal of Molecular Sciences. 2019;**20**(19). DOI: 10.3390/ijms20194873 PMID 31575077

[124] Gilbert L, He X, Farmer P, Boden S, Kozlowski M, Rubin J, et al. Inhibition of osteoblast differentiation by tumor necrosis factor-alpha. Endocrinology. 2000;**141**(11):3956-3964. DOI: 10.1210/endo.141.11.7739 PMID 11089525

[125] Glantschnig H, Fisher JE, Wesolowski G, Rodan GA, Reszka AA. M-CSF, TNFalpha and RANK ligand promote osteoclast survival by signaling through mTOR/S6 kinase. Cell Death and Differentiation. 2003;**10**(10):1165-1177. DOI: 10.1038/sj.cdd.4401285 PMID 14502240

[126] Sihota P, Yadav RN, Poleboina S, Mehandia V, Bhadada SK, Tikoo K, et al. Development of HFD-fed/low-dose STZ-treated female Sprague-Dawley rat model to investigate diabetic bone fragility at different organization levels. JBMR Plus. 2020;**4**(10):e10379. DOI: 10.1002/jbm4.10379 PMID 33103024

[127] Cipriani C, Colangelo L, Santori R, Renella M, Mastrantonio M, Minisola S, et al. The interplay between bone and glucose metabolism. Frontiers in Endocrinology. 2020;**11**:122. DOI: 10.3389/fendo.2020.00122 PMID 32265831

[128] Tanaka K, Yamaguchi T, Kanazawa I, Sugimoto T. Effects of high glucose and advanced glycation end

products on the expressions of sclerostin and RANKL as well as apoptosis in osteocyte-like MLO-Y4-A2 cells. Biochemical and Biophysical Research Communications. 2015;**461**(2):193-199. DOI: 10.1016/j.bbrc.2015.02.091 PMID 25721666

[129] Sanz C, Vazquez P, Blazquez C, Barrio PA, Alvarez MDM, Blazquez E. Signaling VaL and biological effects of glucagon-like peptide 1 on the differentiation of mesenchymal stem cells from human bone marrow. American Journal of Physiology. Endocrinology and Metabolism. 2010;**298**:E634e43. DOI: 10.1152/ajpendo.00460.2009

[130] Shockley KR, Lazarenko OP, Czernik PJ, Rosen CJ, Churchill GA, Lecka-Czernik B. PPARgamma2 nuclear receptor controls multiple regulatory pathways of osteoblast differentiation from marrow mesenchymal stem cells. Journal of Cellular Biochemistry. 2009;**106**(2):232-246. DOI: 10.1002/jcb.21994 PMID 19115254

[131] Schwartz AV. Diabetes mellitus: Does it affect bone? Calcified Tissue International. 2003;**73**(6):515-519. DOI: 10.1007/s00223-003-0023-7 PMID 14517715

[132] Kahn SE, Zinman B, Lachin JM, Haffner SM, Herman WH, Holman RR, et al. Rosiglitazone-associated fractures in type 2 diabetes: An analysis from a diabetes outcome progression trial (ADOPT). Diabetes Care. 2008;**31**(5):845-851. DOI: 10.2337/dc07-2270 PMID 18223031

[133] Loke YK, Singh S, Furberg CD. Long-term use of thiazolidinediones and fractures in type 2 diabetes: A meta-analysis. Canadian Medical Association Journal. 2009;**180**(1):32-39. DOI: 10.1503/cmaj.080486 PMID 19073651

[134] Zhu ZN, Jiang YF, Ding T. Risk of fracture with thiazolidinediones: An updated meta-analysis of randomized clinical trials. Bone. 2014;**68**:115-123. DOI: 10.1016/j.bone.2014.08.010 PMID 25173606

[135] Mancini T, Mazziotti G, Doga M, Carpinteri R, Simetovic N, Vescovi PP, et al. Vertebral fractures in males with type 2 diabetes treated with rosiglitazone. Bone. 2009;**45**(4):784-788. DOI: 10.1016/j.bone.2009.06.006 PMID 19527806

[136] Schwartz AV, Chen H, Ambrosius WT, Sood A, Josse RG, Bonds DE, et al. Effects of TZD use and discontinuation on fracture rates in Accord bone study. The Journal of Clinical Endocrinology and Metabolism. 2015;**100**(11):4059-4066. DOI: 10.1210/jc.2015-1215 PMID 26305617

Chapter 6

Genetic Targets May Be a Promising Future for Osteoporosis

Eiman Mohammad Shahrour

Abstract

The definition, diagnosis and treatment plans for osteoporosis and osteopenia are based on the assessment of BMD by DEXA. However, this method faces many limitations and challenges. The main difficulty is its ability to assess fracture risk. The threshold for evaluating osteoporosis or osteopenia is of high specificity but of low sensitivity. The majority of osteoporotic fractures occur in individuals whose BMD values are above the osteoporotic threshold. These limitations necessitated the search for alternative solutions of better quality, including radiological and genetic ways, and applications with more input risk factors used in fracture risk assessment like FRAX application. Genetic diagnosis of osteoporosis is a real scientific revolution. There are thousands of point mutations implicated in osteoporosis. The future hope is to find a genetic diagnostic method for osteoporosis. This is very necessary because the treatments currently used are to delay the progression of osteoporosis; therefore, an earlier intervention will be effective. In addition, it serves the future prospects for gene therapy for osteoporosis.

Keywords: genetics diagnosis, DEXA, LRP5rs121908669, COL1A2rs72658152, FRAX, osteoporosis

1. Introduction

As defined by the World Health Organization (WHO), osteoporosis is present when BMD is 2.5 SD or more below the average value for young healthy women (a T-score of <−2.5 SD). A second, higher threshold describes "low bone mass" or osteopenia as a T-score that lies between −1 and − 2.5 SD. Osteopenia is a precursor stage of osteoporosis. The difference between the patient's BMD and mean BMD of young females aged in the range of 20–29 years (divided by the standard deviation (SD) of the reference population) yields the T-score; comparing the BMD of a particular age, sex and ethnicity-matched adult reference population is called the Z-score. Treatment plans are linked to specific T-score values [1]. The problem is found in classification, diagnosis, and treatment of osteoporosis [1–5]. Searches are underway for solutions. Among the research directions is the genetic search for solutions to the problems facing DEXA in particular in assessing the condition of the bone and the subsequent selection of the appropriate treatment plan [6–11]. Among the genes selected for study are the COL1A2 gene and LRP5 gene. The COL1A2 gene is chosen because it expresses the collagen protein responsible for bone elasticity by 90% [1]. DEXA completely ignores bone elasticity, as it only measures bone mineral density, which represents only the strength of the bone, not its elasticity. In order to achieve perfect bone, a balance must be achieved between the strength and flexibility of the bone. This is one of the biggest evidences of the inadequacy of DEXA to assess osteoporosis [9]. The COL1A2rs72658152 (located at 7q21.3,G > A1981,OMIM:120160) is chosen specifically because it is inherited in a dominant manner [12, 13] and has proven pathogenicity for postmenopausal osteoporosis. There is no idea how widespread it is in most societies [14]. On the other hand, the LRP5gene is chosen for its relationship to the formation of cortical bone [15], which forms the majority of the femur bone [1], meaning that LRP5 is a very important protein in the pathway for bone formation and construction in the femur region [16, 17] (femur region is one of the DEXA measurement sites, and therefore, it is possible to compare the DEXA results with the genetic results). The LRP5rs121908669 (located at 11q13.2, G > C 511, OMIM: 603506) is chosen because it is inherited predominantly [17], and it has not been genotyped nor in any research worldwide. It is proven to be a form of osteopetrosis. There is no idea how widespread it is in most societies [14]. The spread of applications such as FRAX can be seen as an attempt to solve the problems facing DEXA [11].

2. Methods (LRP5rs121908669 and COL1A2rs72658152)

2.1 Basic methods

As a practical study, the number of participants in the study was 150 women before and after menopause. A DEXA image was taken for all participants. The participants were divided into three groups according to the current classification of the World Health Organization related to T-Score (normal, osteopenia, and osteoporosis). Normal BMD female participants formed as the control group. Cases of mild osteogenesis imperfecta were diagnosed by a specialist in arthritis and rheumatology according to Sillence standards. The participants were classified into two groups who had symptoms of mild osteogenesis or not. A questionnaire was collected from all the participants, which included information about age, height, weight, body mass index, age of onset of bone complaint, age of the beginning of the menstrual cycle, age of the end of the menstrual cycle, number of children, family history, type of work. Women suffering from hypertension, diabetes, osteomalacia, surgical menopause, and cancer were excluded. Blood samples were collected from the participants on EDTA tubes. PCR, RFLP, DNA sequencing were performed for all samples for two SNPs (LRP5rs121908669 and COL1A2rs72658152). The work was done in accordance with the ethics of scientific research at Tishreen University, Syria.

2.2 Statistical study

The participants were distributed according to the classification of the World Health Organization into three groups (normal, osteopenia, osteoporosis). The distribution of participants was according to the diagnosis of mild osteogenesis imperfecta into two groups (yes/no). The participants with mild osteogenesis imperfecta were distributed according to the groups (normal, osteopenia, osteoporosis) into three groups. For the genetic results, no COL1A2rs72658152 appeared in any participant, so an attempt was made to study the pathological association which it is caused by this SNP. The genotypes of the LRP5rs121908669 were distributed according to three groups (GG, CC, and GC).

Several statistical applications (binary logistic regression test, Related-Samples McNemar Change Test, chi-square test) were used to study the relationship between mild osteogenesis imperfecta and postmenopausal osteoporosis and osteopenia. Chi-square test was used to study the association between carrier of (mild osteogenesis imperfecta—postmenopausal osteoporosis, osteopenia) and clinical data (height, age of onset of bone complaint). Statistical applications (Related-Samples McNemar Change Test, chi-square Test, Odd Ratio test) were used to study the relationship of genotypes of LRP5rs121908669 with femur T-score and the relationship of genotypes of LRP5rs121908669 with lumbar T-score. Chi-square test, Odd Ratio test, and likelihood ratio test were used to study the relationship of genotypes of LRP5rs121908669 with clinical data (body mass index).

3. Results

3.1 Data values for the participants

All the data of the studied participants are presented in the **Table 1**. It was obtained by the personal question of the participants and by referring to their files in the hospital and by clinical and radiological diagnosis. It includes data about the age of onset of the bone complaint, the age of the onset and end of the menstrual cycle, the number of children, the family history of the bone complaint, the history of bone

Variable	Case
Total number	150
Age	60(40, 80)
Age of beginning of menstrual	14(11, 17)
Age of end of menstrual	50.5(46, 55)
Weight	69.5(40,99)
Height	165(150,180)
BMI	29.69(17.99, 41.4)
Hearing features(YES/NO)	31/119
Dental features(YES/NO)	68/82
Data on fractures(YES/NO)	85/65
History of family orthopedic complaint(YES/NO)	56/94
Clinical history of bone complaint(YES/NO)	139/11
L2-L4(lumbar) Z-score	(−4.1, 3.1)
L2-L4 (lumbar)T-score	(−5.6, 1.2)
Femur Z-score	(−1.9, 1.1)
Femur T-score	(−2.2, 1.1)
Normal(T-score ≥ 1)	74
Osteopenia (−2.5) < T-score < (−1)	48

Variable	Case
Osteoporosis T-score ≤ (−2.5)	28
Total (normal, osteopenia, osteoporosis)	150
Normal, mild OI	58/99
Osteopenia, mild OI	27/99
Osteoporosis, mild OI	14 /99
Total, mild OI	99/150

Table 1.
Data of the participants in the study (the results of the clinical diagnosis, the results of the radiological diagnosis, the data of the total questionnaire).

pain in the participants, measurements of height and weight, body mass index BMI (kg/m2), history of fractures, classification of cases according to T-score according to the World Health Organization. The number of cases of mild osteogenesis imperfecta was according to the evaluation of specialists. Serum calcium and phosphorus concentrations were normal for all selected participants (**Figures 1–4**).

3.2 Genetic findings

3.2.1 Results of RFLP and DNA sequencing tests for both LRP5rs121908669 and COL1A2rs72658152

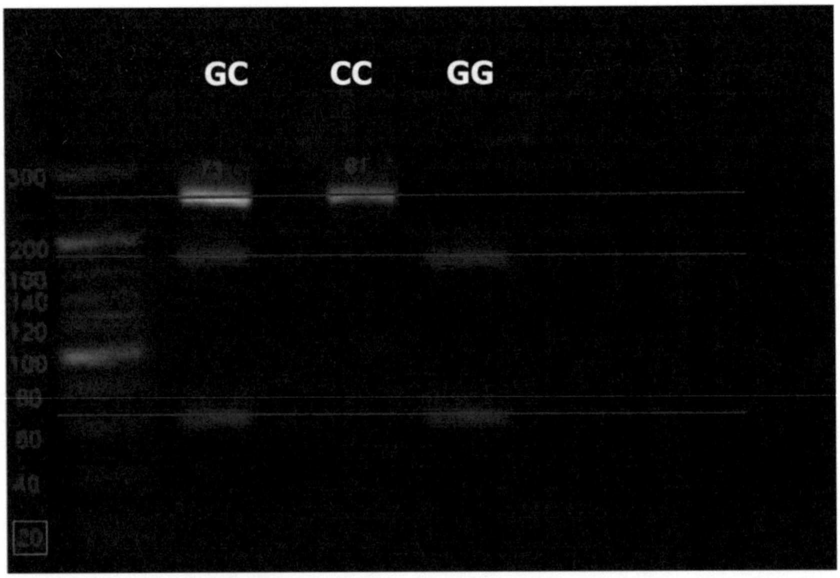

Figure 1.
LRP5rs121908669, results of migration of PCR products digested with BfiI enzyme (RFLP technique) on agarose gel, DNA ruler 20 bp, genotype CC (259) bp, genotype GG (bp67, 192 bp), genotype GC (bp67, bp192, bp259).

DOI: http://dx.doi.org/10.5772/intechopen.110336

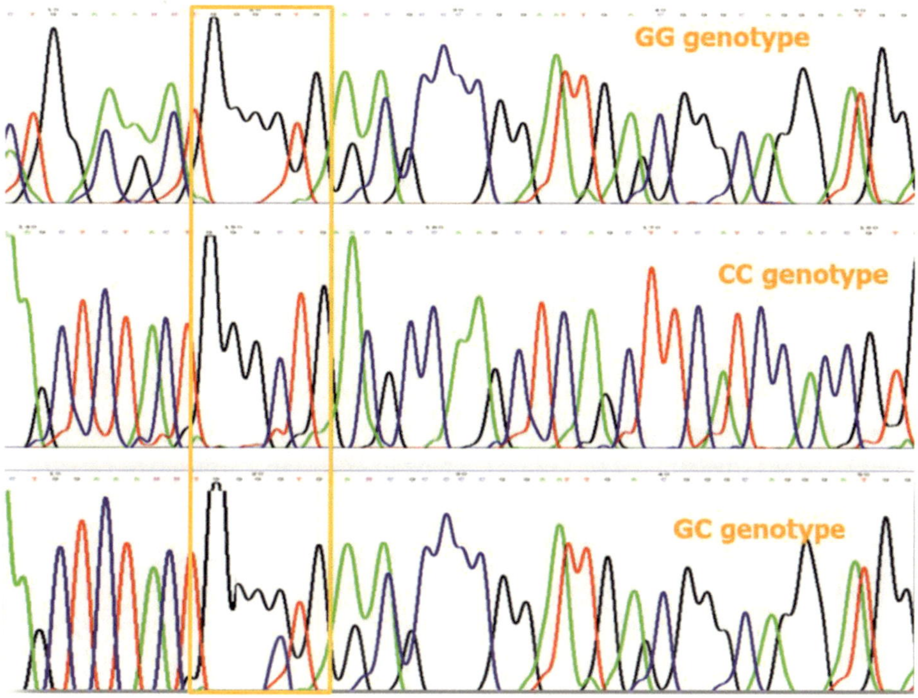

Figure 2.
DNA sequencing results for the three genotypes, LRP5rs121908669. GG(GGGGT), CC(GGGCT), GC(GGG(G/C)T).

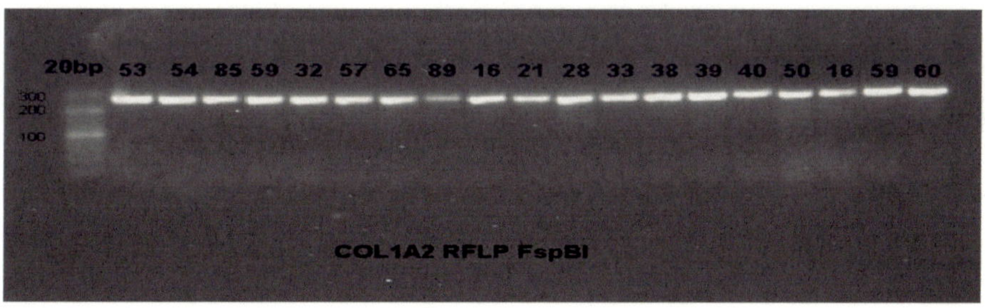

Figure 3.
Migration results of COL1A2rs72658152, COL1A2rs72658152, digested with FspBI enzyme (RFLP technique) on agarose gel, DNA ruler 20 bp, genotype GG (255) bp.

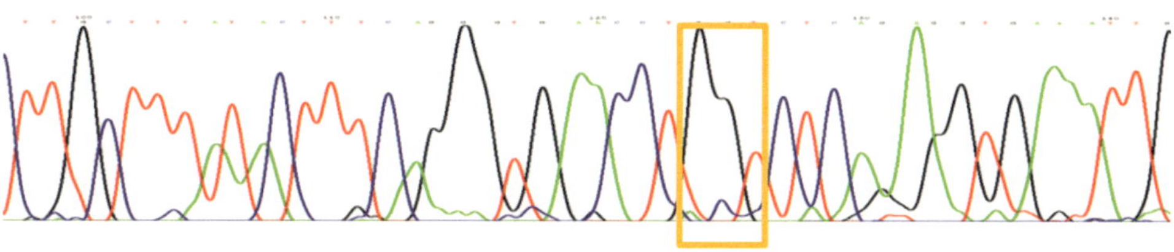

Figure 4.
COL1A2rs72658152, DNA sequence results, negative results, normal genotype is (CTGG) GG(255 bp).

3.2.2 The frequency of SNPs LRP5rs121908669 and COL1A2rs72658152

The results of the PCR-RFLP and DNA sequencing tests for each of the two mutations were as shown in **Table 2**:

LRP5rs121908669: This study is the first of its kind in Syria as a genetic study related to LRP5, and it is the first study of its kind in the world regarding the genotyping of LRP5rs121908669.

It was found that there were 52 (34.66%) mutant cases (CC, GC) and 98 (65.33%) normal cases (GG). The mutant cases were distributed to 20 (13.3%) homozygous genotype CC and 32 (21.33%) heterozygous genotype GC.

The mutated cases (CC, GC) were distributed among low BMD cases (osteopenia, osteoporosis) and the control group (normal BMD cases) to ((19.51%) (16) (36) (52.94%)), respectively.

The proportion and number of cases of the homozygous genotype CC included 20 (13.3%). It was distributed to 5 (25%) versus 15 (75%) in the group of cases of low bone mineral density (osteopenia and osteoporosis) and the control group (cases with normal bone mineral density), respectively.

The percentage and number of cases of heterozygous genotype GC 21.33 (32%) were included. It was distributed to 11 (34.37%) compared to 21 (65.62%) in the group of BMD cases (osteopenia and osteoporosis) and the control group, respectively.

The proportion and number of cases of the normal genotype included GG 98 (65.33%). It was distributed among 66 (67.34%) compared to 32 (32.65%) in the group of carriers of low bone mineral density (osteopenia and osteoporosis) and the control group, respectively.

The CC and GC genotypes are associated with cases with normal BMD values in higher percentages than cases with low BMD values. This does not agree with the idea of an association between LRP5rs121908669 and cases with high bone mineral density—regardless of genotype—as reported in a Belgian study by Liesbeth Van Wesenbeeck et al. 2002 [17]. But it reinforces the studies indicating the existence of ethnic differences in the expression of the mutations themselves [3] (Europe, Middle East).

The association of CC and GC genotypes with low BMD cannot be neglected in some cases, contrary to the results of the Belgian study by Liesbeth Van Wesenbeeck et al. [17]. These cases may be attributed to the presence of protective factors (genetic or environmental factors) against the expression of the mutant-fixing gene (association with high BMD) or the presence of an interaction between the genes. Thus, the difference in the genetic expression of the mutation genotypes may be reflected in their effect on bone mineral density [18].

%	Number	Genotyping
64.66	97	GG/ LRP5rs121908669
13.3	20	CC/ LRP5rs121908669
22	33	GC/ LRP5rs121908669
No positive case was recorded, COL11A2G661S		

Table 2.
Frequency of genotypes of the two studied mutations.

Not all carriers of CC and GC genotypes with various BMD states show clinical symptoms of autosomal dominant osteopetrosis1 ADO1, even though it is a proven pathogenic mutation of an ADO1. This contradicts the study by Liesbeth Van Wesenbeeck et al., 2002 [17]. It has been reported that LRP5rs121908669 is associated with ADO1 which is sometimes associated with generalized bone pain and hearing loss but certainly not associated with fractures [15–17, 19]. These differences in gene expression at the BMD level may be explained by race-related factors or genetic or environmental factors affecting gene expression [15].

It is possible that the clinical features in these studied participants with bone mineral density above +1 for LRP5rs121908669 are related to HBM high bone mineral more than to ADO1, and more clinical genetic studies are needed for further understanding. High bone mineral mass (HBM) and ADOI are diseases from the osteoporosis group. The radiological features are strikingly similar but HBM patients clinically have no complaints and are completely asymptomatic [20, 21] while at least some ADOI patients present with severe pain [20, 22]. ADOI is the only type of osteopetrosis that is not associated with an increased fracture rate but HBM is associated with an increased fracture rate [17, 19, 20].

The prevalence of cases carrying GC and CC genotypes is 52 (34.66%). It is a big percentage. This contradicts studies indicating that it is a rare mutation [15].

COL1A2rs72658152: No positive result was recorded. Therefore, it was directed to study the relationship of mild OI with low bone mineral density after menopause. This conjugation has been shown to be due to the COL1A2rs72658152 mutation, according to the NCBI.

3.2.3 The relationship of genotypes of LRP5rs121908669 with lumbar bone mineral density

The results of Odd ratio test, chi-square test, and Related-Samples McNemar Change Test which were applied to study the relationship between the genotypes of LRP5rs121908669 and lumbar BMD are shown in **Table 3**.

Osteopenia increases and osteoporosis decreases in lumbar bone (L1-L4) in carriers of the CC genotype. The appearance of normal BMD increases and osteoporosis decreases in lumbar bone in carriers of the GG genotype. The appearance of low lumbar bone mineral density (osteopenia) increases among carriers of the GC genotype.

It is noted that the genotypes of LRP5rs121908669 have an effect on the bone mineral density in the lumbar region, but the question could be "does it have an effect as important as the effect of hormonal factors on the bone mineral density in the lumbar region?"

In order to answer this question, a comparative study must be made between the statistical results in **Table 3** specially between the participants who carry the CC and GC genotypes and suffer from low bone mineral density in the lumbar region and the participants who carry the GG genotype and suffer from low of bone mineral density in the lumbar region. Genetic patterns have greater influence than hormonal factors, knowing that all participants are women per and after menopause. It is preferable to measure the estrogen hormone in the participants and compare its concentrations with the appearing of genotypes and study the effect on the bone mineral density in the lumbar region.

This study is the first of its kind in the world. There is only one Belgian study that revealed this mutation by chance and it has not studied afterward, but in that study

Genotypes	N.	BMD	McNemar%	OR	CI	Chi-square	P
GG	1	Normal	51.5	1.4000	1.092–1.794	6.302	0.012
	2	Osteopenia	0	0.495	0.268–0.913	5.919	0.015
	3	Osteoporosis	0	0.813	0.263–2.516	0.130	0.719
CC	1	Normal	0	0.712	0.546–0.928	3.846	0.05
	2	Osteopenia	1	3.462	0.910–13.165	7.731	0.005
	3	Osteoporosis	26.5	0.846	0.202–3.540	0.204	0.651
GC	1	Normal	0	0.822	0.625–1.081	1.658	0.198
	2	Osteopenia	10.4	1.375	0.714–2.648	0.989	0.320
	3	Osteoporosis	3	1.551	0.362–6.655	0.363	0.547

Table 3.
Results of the statistical relationships that study the relationship between lumbar T-score values and genotypes of LRP5rs121908669.

it was associated with an increase in bone mineral density without specifying the genotype, and this is contrary to the results of this research.

3.2.4 The relationship of genotypes of LRP5rs121908669 with femoral bone mineral density

The results of the odd ratio test, chi-square test, and McNemar test are shown in **Table 4**. These applications were applied to study the relationship between the genotypes of LRP5rs121908669 and femoral bone mineral density.

The appearance of low femoral bone mineral density (osteopenia, osteoporosis) increases and normal femoral bone mineral density decreases in carriers of the CC genotype. The appearance of normal femoral bone mineral density increases and the appearance of low femoral bone mineral density (osteopenia) decreases in carriers of the GG genotype. The appearance of low lumbar bone mineral density (osteopenia, osteoporosis) increases among carriers of the GC genotype.

It is noted that the genotypes of the mutation have effects on the bone mineral density in the femoral region and with a stronger effect than their effect on the bone mineral density in the lumbar region.

It can be explained that the LRP5rs121908669 mutation is one of the mutations of the LRP5 gene that encodes the LRP5 protein which is a part of the WNT pathway. WNT pathway is one of the most important pathways involved in the formation of cortical bone [19]. The cortical bone makes up the majority of the femur bone. Thus, the emergence of the greater effect of the genotypes of LRP5rs121908669 can be explained in the femur than in the lumbar bone, since the lumbar bone is formed mostly from spongy bone [1]. For more clarification, each of the femur and the lumbar bone consists of cortical bone and cancellous bone, but the proportion of cortical bone predominates in the femur, and the proportion of cancellous bone predominates in the lumbar bone [1].

This study is the first of its kind in the world. There is only one Belgian study that revealed this mutation by chance and it was not studied afterward, but in that study it was associated with an increase in bone mineral density without specifying the

Genotype	Groups	McNemar %	OR(Odds ratio)	CI(95% confidence interval)	Chi-square	P
GG	Normal	2.5	1.830	1.344–2.493	13.750	0.000
	Osteopenia	0	0.366	0.185–0.723	10.765	0.001
	Osteoporosis	0	0.732	0.346–1.547	0.689	0.407
CC	Normal	0	0.558	0.417–0.746	8.683	0.003
	Osteopenia	1	7.231	1.056–49.516	7.731	0.005
	Osteoporosis	28	1.282	0.426–3.856	0.204	0.651
GC	Normal	0	0.712	0.514–0.986	3.463	0.063
	Osteopenia	8.7	1.652	0.818–3.334	2.263	0.133
	Osteoporosis	57.5	1.297	0.535–3.149	0.344	0.557

Table 4.
Results of the statistical relationships that study the relationship between femur T-score values and genotypes of LRP5rs121908669.

genotype, and this is contrary to the results of this research. However, it seems that it will receive wide attention after it was added to the lists of genetic laboratory kits for one of the major American international companies to manufacture laboratory kits for genetic analysis in 2022.

3.2.5 Correlation of the genotypes of LRP5rs121908669 with body mass index BMI

According to the results of the odd ratio test, the logistic regression test, and the chi-square test which are presented in **Table 5**, the following results are obtained:

The appearance of the genotype GG increases in women who are carriers of BMI (obese, over obesity) and decreases in women with normal BMI. In other words, the characteristics of the carriers of the GG genotype in terms of BMI may be (obese or over obesity) and they cannot have a normal BMI. The GG genotype is a predisposing factor for obesity and over obesity, and a protective factor against reaching normal BMI.

The appearance of the CC genotype decreases in women who are carriers of BMI (overweight, over obesity). In other words, carriers of the CC genotype cannot be with BMI (overweight or over obesity). The CC genotype is a protective factor against BMI (over obesity and overweight).

The appearance of the genotype GC decreases in women who are carriers of BMI (normal, obese). In other words, the characteristics of GC genotype carriers may be in terms of BMI (normal, obesity). The GC genotype is a protective factor against obesity and a protective factor against reaching BMI (normal weight). This relationship is being studied for the first time globally.

Genotypes	N.	BMI groups	B/ OR	CI / P	Exp(B)/ OR	Chi-square	P
GG	1	<18.5	.915	0.173–4.832	.915	0.011	0.917
	2	[18.5–24.9]	.179	0.68–.471	.179	19.675	0.000
	3	[25–29.9]	.707	0.452–1.107	.707	2.507	0.113
	4	[30–34.9]	5.491	.2.499–12.064	5.491	23.707	0.000
	5	≥35	16.472	2.145–126.510	16.472	14.013	0.000
CC	1	<18.5	−8.244	.118	.000	0.060	0.806
	2	[18.5–24.9]	−8.374	.064	.000	6.868	0.009
	3	[25–29.9]	−4.712	0.016	.009	0.307	0.579
	4	[30–34.9]	−3.280	.082	.038	1.952	0.162
	5	≥35	.154	.049–.484	.154	12.466	0.000
GC	1	<18.5	−2.271	.455	.103	0.104	0.748
	2	[18.5–24.9]	−2.802	.300	.061	8.808	0.003
	3	[25–29.9]	−1.288	.421	.276	1.883	0.170
	4	[30–34.9]	.650	.673	1.915	19.997	0.000
	5	≥35	.423	.127–1.411	.423	2.023	0.155

Table 5.
Results of the statistical relationships that study the relationship between BMI and genotypes of LRP5rs121908669.

3.2.6 The relationship of mild osteogenesis imperfecta OI with low bone mineral density after menopause

According to chi-square test results, the appearance of mild OI can be associated with all cases of bone mineral density in the lumbar position (the appearance of this association decreases with normal bone mineral density and increases in the rest of the cases). The appearance of mild OI is associated with cases of low bone mineral density in the lumbar position or the femoral position.

According to Related-Samples McNemar Change Test results, there is an association between the appearance of mild OI and normal bone mineral density in the femoral position (an inverse relationship).

No participant showed the studied collagen mutation (COL1A2rs72658152). However, there is a study indicating that COL1A2rs72658152 mutation is present in women who are carriers of symptoms of mild OI and at the same time they have osteoporosis or osteopenia after menopause (low bone mineral density). This study was a solution to one of the drawbacks of Loretta's study [12, 23], the number of participants is larger. In fact, there is an association between carriers of symptoms of mild osteogenesis imperfecta and postmenopausal osteoporosis. What led to thinking that osteoporosis after menopause is not a primary osteoporosis [24] due to the presence of genetic causes for it and because it is a consequence of genetic diseases, as mild OI [23]. Mild OI is a genetic disease whose symptoms are very mild, it does not have serious clinical manifestations, but after menopause the condition develops into postmenopausal osteoporosis. This summary is explained statistically according

Mild OI with		McNemar%	B/OR	P/CI	OR/ Exp(B)	Chi-Square Values	P
Lumbar	Normal	0.1	0.414	0.252–0.678	0.414	17.575	0.000
	Osteopenia	0	1.941	1.233–3.056	1.941	8.053	0.005
	Osteoporosis	0	1.941	1.004–3.753	1.941	3.927	0.048
	Low BMD	2.7	1.941	1.443–2.612	1.941	17.575	0.000
Femur	Normal	32.8	0.923	0.694–1.227	0.923	0.317	0.573
	Osteopenia	0	1.100	0.674–1.795	1.100	0.144	0.705
	Osteoporosis	0	1.213	0.418–3.520	1.213	0.126	0.722
	Low BMD	0	1.124	0.752–1.680	1.124	0.317	0.573
Lumbar/femur	Low BMD	10.8	1.758	0.000	5.800	20.635	0.000

Table 6.
Results of related-samples McNemar change test, odd ratio tests, chi-square tests to evaluate the relationship of mild OI to bones statues (normal, osteopenia, osteoporosis).

		B	P/CI	Exp(B)	Chi-square	P
Carriers of (Mild OI, low BMD)	[40–59] weight	2.868	.030	17.597	1.645	0.200
	[60–79] weight	.934	.317	2.544	0.724	0.395
	[80–99] weight	1.686	.092	5.397	9.865	0.352
	[150–159] height	.069	.951	1.071	1.165	0.280
	[160–169] height	.239	.816	1.270	0.467	0.494
	[170–179] height	4.138	1.018–16.817	4.138	5.193	0.023
	18.5 ≥ BMI	−2.590	.204	.075	0.358	0.550
	[18.5–24.9] BMI	−1.534	.294	.216	0.078	0.780
	≥35 BMI	1.505	0.333–6.793	1.505	0.290	0.590
	[40–50] age	.501	.712	1.650	1.477	0.224
	[50–60] age	1.668	.211	5.299	0.358	0.550
	[60–70] age	2.839	.033	17.107	6.253	0.012
	70–80 age	6.771	0.934–49.107	6.771	5.335	0.021
	[11] first of Menstrual	1.182	.229	3.262	2.955	0.086
	[12] first of Menstrual	−.457	.653	.633	2.713	0.100
	[13] first of Menstrual	.884	.349	2.419	2.212	0.137
	[14] first of Menstrual	.467	.614	1.595	0.016	0.898
	[15] first of Menstrual	2.257	0.702–7.259	2.257	2.093	0.148
	[40–42] end of Menstrual	.237	.763	1.268	0.329	0.566
	[43–45] end of Menstrual	.677	.391	1.967	0.401	0.527
	[46–48] end of Menstrual	−.040	.956	.961	1.961	0.161
	[49–51] end of Menstrual	−1.103	.144	.332	2.035	0.154

Table 7.
Results of the statistical relationships that study the relationship between carriers of {mild OI with low bone mineral density in the lumbar or femoral position (osteoporosis or osteopenia) after menopause} with the data of the participants.

to **Table 6** containing the results of statistical studies (McNemar Change Test, Odd Ratio test, chi-square test) which were applied to study the relationship between mild osteogenesis imperfecta and femoral and/or lumbar bone mineral density in all its forms (normal/osteoporosis/osteopenia) (**Figure 4**, **Table 7**).

3.2.7 The relationship of carriers (mild OI with low bone mineral density in the lumbar or femoral position (osteoporosis or osteopenia) after menopause with the data of the participants)

According to chi-square test results, this comorbidity (mild OI, low BMD) increases in tall women 170–179, and women who show a peak in bone pain complaints at the age of 60–80.

According to Related-Samples McNemar Change Test results, the emergence of bone pain complaints at the age of 60–70 is considered a risk factor for the emergence of this comorbidity.

3.2.8 Results of genetic test (LRP5rs121908669, COL1A2rs72658152) and FRAX vs. DEXA results (1: 0) as a case report

By comparing the genetic results of the two studied SNPs and the results of the application of FRAX and the results of DEXA for two clinical cases, there was agreement between the results of FRAX and the genetic results without their compatibility with the results of DEXA. The first case is 75 years old. Results of DEXA are osteopenia in the femur bone and normal in the lumbar bone. Results of FRAX are major osteoporotic 20, hip fracture 9.7. Therapeutic intervention is decided by the doctor. The genotype of LRP5rs121908669 is GC. There is not COL1A2rs72658152.

The second case is 45 years old. Results of DEXA are osteoporosis in the femur and lumbar bones. Results of FRAX are major osteoporotic 0.4, hip fracture 0. Therapeutic intervention is not needed as the doctor's opinion. The genotype of LRP5rs121908669 is GG. There is not COL1A2rs72658152.

The results of DEXA conflict with the opinion of the specialist, while the results of FRAX agree with the opinion of the specialist. Genotypes of LRP5rs121908669 agree with the FRAX results, but vary with DEXA radiographic findings. FRAX is not a diagnosis tool. FRAX is for predicting fractures. But logically, it can be used in the validity of the current diagnosis, especially with the existence of the problem of the approved criteria (DEXA).

When a case is diagnosed as osteoporosis and requires therapeutic intervention according to the approved criteria, then FRAX is applied and the results are not predictive of fractures. Illogical something has been facing (case2). Or, when the case is not diagnosed as osteoporosis FRAX is applied and the results are predictive of fractures. In fact, this case requires therapeutic intervention, although it does not conform to the standards of the World Health Organization (case1). This requires more studies for comparing the results of FRAX with the results of DEXA on a larger scale of samples. This comparison is the first of its kind in the world.

4. Conclusion

- The results of the FRAX application show consistency with the results of the LRP5rs121908669 genetic diagnosis and the diagnosis based on the opinion of the clinician more than the compatibility of the results of DEXA with the genetic and clinical diagnosis. The World Health Organization has been contacted in this regard. Syrian Tishreen University is the first discoverer of this idea. This also strengthens the link between Tishreen University research and community services. This is a research service on a global and local level. It can be suggested to add the genetic factor especially LRP5rs121908669 to the application of FRAX. Syrian Tishreen University is proposed this idea for the first time in the world to prof. John Kanis.

- The association between mild osteogenesis imperfecta and postmenopausal osteoporosis is significant statistically. But there seems to be another reason for this association than COL1A2rs72658152. Postmenopausal osteoporosis is not primary osteoporosis. The World Health Organization has been contacted in this regard, and Syrian Tishreen University is the first discoverer of this idea.

- -The genotyping of LRP5rs121908669 is the first worldwide genotyping. Statistical studies showed the relationship of mutant genotypes (GC, CC) to osteopenia and osteoporosis in the lumbar region and the femur region, and the relationship of the normal genotype (GG) to normal bone mineral density in the lumbar region and the femur region. This is a world class search service. Syrian Tishreen University is the first discoverer of this idea.

- Cases of low bone mineral density in the femur region before the lumbar region in some postmenopausal women can be explained by the presence of the CC or GC mutant types of LRP5rs121908669 because of their clear effect on cortical bone (the relationship of LRP5 and cortical bone). This is a logical explanation for a realistic problem that was not previously explained, and question marks are always placed on it. This is a research service on a global and local level. Syrian Tishreen University is the first discoverer of this idea.

- It is obvious that cases of low of bone mineral density in the lumbar region are due to hormonal reasons (the relationship of spongy bone and hormonal factors), but it was found that there is a relationship to the genetic factor as well. This is a world class search service.

- The existence of specific morphological characteristics of the carriers of the LRP5rs121908669genotypes, and this reinforces the idea of morphological genetics to assist in the final clinical diagnosis. This is a world class search service. Syrian Tishreen University is the first discoverer of this idea.

- The current treatment plan should be based on drugs that reduce bone resorption or increase bone formation, according to the effect of the genetic variation and according to the signaling pathway to which it belongs. In addition to the idea of developing a targeted gene therapy for osteoporosis. It can be included in future prospects for the treatment of osteoporosis.

Acknowledgements

We express gratitude to our Research Team, Faculty of Pharmacy of Tishreen University and Atomic Energy Commission of SYRIA (AECS) for their support. Big thanks to Prof. Jean-Yves Reginster and Prof. Didier Hans and Prof. Stephen Dolye and Prof. Emmanuel Maheu for their support. Big Thanks for Prof. Nizar Mirali, Prof. Walid Al-achkar, Prof. Abd Alrazak Hassan and Prof. Haissam Yazigi. Big thanks to Sir. Ayman albloj, Ms. Sarah Reslan. Big thanks to Mss. Maison younes. Big thanks to Pharmacist. Mohammed Kbeili to help with statistical analysis. Huge thanks to the presidency of Tishreen University and the National Union of Syrian Students, Tishreen University, for the great support provided to me.

Conflict of interests

Eiman shahrour declares that they have no conflict of interest.

Funding

This research received no external funding.

Declaration

I confirm that this work is a part of an approved PhD thesis which was approved by university board's decision No.1698 of 05/02/2019, and this work is an original and has not been published elsewhere, nor is it currently under consideration for publication elsewhere.

Ethical approval statement

The work was approved by the Ethics Committee in Syrian Ministry of Higher Education and written informed consent was obtained from all the participants according to the Declaration of Helsinki. The ethical approval for this clinical case study has the number N.1698 of 05/02/2019.

Informed consent statement

"Informed consent" was obtained from all subjects involved in the study. Written informed consent has been obtained from the patient(s) to publish this paper.

Data availability statement

The data that support the findings of this study are available from the corresponding author (Eiman M. Shahrour), upon reasonable request. All relevant material is included in this publication.

Supplemental material

Supplemental material referenced in this chapter can be downloaded at: https://bit.ly/3l1KEOV

Author details

Eiman Mohammad Shahrour
Faculty of Pharmacy, Department of Biochemistry and Microbiology, Ministry of High Education, Tishreen University, Lattakia, Syria

*Address all correspondence to: eimanm.shahrour@gmail.com